Plate I. Troglocambarus maclanei Hobbs

UNIVERSITY OF FLORIDA PUBLICATION

NOVEMBER, 1942

Vol. III   BIOLOGICAL SCIENCE SERIES   No. 2

THE CRAYFISHES OF FLORIDA

BY

HORTON H. HOBBS, JR.

Published by the University of Florida under the auspices of the
Committee on University Publications

QL444
.D3H64

4119

TABLE OF CONTENTS

	PAGE
THE CRAYFISHES OF FLORIDA	1
Acknowledgments	1
Material Studied	3
Methods of Collecting	4
Field Notes	7
Methods of Preservation and Study	7
Previous Work on the Crayfishes of Florida	8
FACTORS AFFECTING THE DISTRIBUTION OF CRAYFISHES IN FLORIDA	9
Florida as a Faunal Region	9
Derivation of the Fauna	10
Endemism	12
Principal Migration Routes	13
Barriers to Crayfish Migration	14
Ecology	16
Ecological Habits and Habitats	20
THE SUBFAMILY CAMBARINAE	23
The Evaluation of Taxonomic Characters in the Cambarinae	23
The Chief Characters Used in Keys and Descriptions	25
Key to the Florida Crayfishes	28
GENUS PROCAMBARUS	33
Barbatus Section	33
Barbatus Group	35
Barbatus Subgroup	38
Procambarus barbatus	39
Procambarus pubischelae	41
Procambarus escambiensis	46
Procambarus econfinae	49
Procambarus latipleurum	52
Procambarus apalachicolae	55
Procambarus rathbunae	59

TABLE OF CONTENTS—Continued

	PAGE
Shermani Subgroup	60
Procambarus shermani	61
Kilbyi Subgroup	64
Procambarus kilbyi	64
Hubbelli Subgroup	67
Procambarus hubbelli	67
Alleni Group	69
Procambarus alleni	69
Advena Section	73
Advena Group	73
Procambarus advena	75
Procambarus geodytes	80
Procambarus pygmaeus	83
Rogersi Group	88
Procambarus rogersi rogersi	89
Procambarus rogersi ochlocknensis	89
Procambarus rogersi campestris	90
Acherontis Section	91
Procambarus acherontis	91
Blandingii Section	93
Blandingii Group	93
Blandingii Subgroup	93
Procambarus blandingii acutus	94
Procambarus bivittatus	96
Clarkii Subgroup	98
Procambarus okaloosae	100
Procambarus paeninsulanus	104
Evermanni Subgroup	107
Procambarus evermanni	107
Fallax Subgroup	111
Procambarus fallax	111
Procambarus leonensis	114
Procambarus pycnogonopodus	117

TABLE OF CONTENTS—Continued

	PAGE
Spiculifer Group	119
Procambarus spiculifer	119
Procambarus versutus	126
Pictus Group	129
Pictus Subgroup	129
Procambarus pictus	130
Procambarus youngi	131
Lucifugus Subgroup	134
Procambarus lucifugus lucifugus	134
Procambarus lucifugus alachua	136
Procambarus pallidus	139
Seminolae Subgroup	142
Procambarus seminolae	142
GENUS TROGLOCAMBARUS	146
Troglocambarus maclanei	146
GENUS CAMBARELLUS	149
Cambarellus schmitti	149
GENUS ORCONECTES	153
Subgenus Faxonella	154
Orconectes clypeata	154
GENUS CAMBARUS	156
Bartonii Section	157
Cambarus latimanus	158
Cambarus floridanus	161
Cambarus cryptodytes	162
Diogenes Section	164
Cambarus diogenes	164
Cambarus byersi	167
Cambarus (species incertis)	168
A COUNTY LIST OF THE FLORIDA CRAYFISHES	170
LITERATURE CITED	172
INDEX	177

TABLE OF CONTENTS—Continued

TEXT FIGURES

	PAGE
a—Specimens of Field and Species Catalogue Cards	6
b—Terminology and Methods of Taking Measurements	24
c—Hypothetical First Pleopod of Male	26

MAPS

1—County Map of Florida Showing Areas in Which Field Work Has Been Done	2
2—Continuous North-South Extents of Well Drained Soils That Appear to Act as Barriers to Crayfish Migration	14
3—Migrations of the Barbatus Group	34
4—Distribution of the Barbatus Group	36
5—Distribution of the Advena Section, *Procambarus alleni*, and *Procambarus blandingii acutus*	74
6—Distribution of the Clarkii Subgroup	99
7—Distribution of the Evermanni and Fallax Subgroups	110
8—Distribution of the Spiculifer Group and *Procambarus acherontis*	120
9—Distribution of the Pictus Group and *Procambarus bivittatus*	130
10—Distribution of the Genera *Troglocambarus*, *Cambarellus*, and *Orconectes* in Florida	148
11—Distribution of the Genus *Cambarus* in Florida	156

PLATES

I—Frontispiece—*Troglocambarus maclanei*

II—*Procambarus barbatus, P. pubischelae, P. escambiensis, P. econfinae*

III—*Procambarus latipleurum, P. apalachicolae, P. rathbunae, P. shermani*

IV—*Procambarus kilbyi, P. hubbelli, P. alleni, P. advena*

V—*Procambarus geodytes, P. pygmaeus, P. rogersi rogersi, P. rogersi ochlocknensis*

VI—*Procambarus rogersi campestris, P. acherontis, P. blandingii acutus, P. bivittatus*

VII—*Procambarus okaloosae, P. paeninsulanus, P. evermanni, P. fallax*

VIII—*Procambarus leonensis, P. pycnogonopodus, P. spiculifer, P. versutus*

IX—*Procambarus pictus, P. youngi, P. lucifugus lucifugus, P. lucifugus alachua*

TABLE OF CONTENTS—Continued

X—*Procambarus pallidus, P. seminolae, Troglocambarus maclanei, Cambarellus schmitti*

XI—*Orconectes clypeata, Cambarus latimanus, C. floridanus, C. cryptodytes*

XII—*Cambarus diogenes, C. byersi, C. species incertis*

XIII—*Procambarus pubischelae*

XIV—*Procambarus escambiensis*

XV—*Procambarus econfinae*

XVI—*Procambarus latipleurum*

XVII—*Procambarus apalachicolae*

XVIII—*Procambarus shermani*

XIX—*Procambarus geodytes*

XX—*Procambarus pygmaeus, Cambarellus schmitti*

XXI—*Procambarus bivittatus*

XXII—*Procambarus okaloosae*

XXIII—*Procambarus youngi*

XXIV—*Procambarus seminolae*

THE CRAYFISHES OF FLORIDA

Although the crayfishes of the subfamily Cambarinae are among the most familiar of all invertebrate animals, many of the species and the group as a whole present a number of interesting biological problems. Wholly Nearctic in origin and distribution, the subfamily now occupies a large area in North America east of the Continental Divide in spite of the low vagility of most of the species and a marked tendency to be limited by topographical and ecological barriers. Here in the vast extent of latitude, topography, and climate between south-central Canada and Guatemala the crayfishes of this group have encountered and shown numerous adaptations to a wide range of ecological conditions.

Perhaps it is for these reasons that the crayfishes have attracted a number of able students, who have not only laid a good basis for taxonomic and geographic work but, at least in the case of Ortmann, utilized the group for classical researches in geographic distribution. Most of this work, however, has dealt with regions other than the southeastern coastal plain. Here, except for the pioneer work of LeConte, the group has been neglected, and there is little evidence that any of the older workers suspected that this region and especially Florida would prove to be exceptionally rich in species and present so many striking instances of recent migration, speciation, and adaptation to specialized habitats.

My own work, begun ten years ago, was at first visualized as the comparatively simple task of locating some eight or ten species that had been reported for Florida, learning to recognize them in the field, and determining their local geographic and ecological ranges. It soon became apparent, however, that Florida did not present a simple problem. As the collections grew in numbers of specimens and in the regions and habitats represented, it became evident that I had to deal with an almost uniquely rich and varied crayfish fauna. Today the Florida list has grown to 42 species and subspecies, the largest list for any state in the Union, and the details of their occurrence present a wealth of data on the comparatively recent spread of the stock into the region, on its encounters with ecological highways and barriers to migration, and the consequent development of a high proportion of endemic races and species.

Acknowledgments

The present investigation has been carried out with the active encouragement of Professor J. Speed Rogers. His suggestions have been invaluable in working out the details of the problem, and I have gone to him freely for criticism of the manuscript. For this aid, for the loan of equipment, and for the personal interest he has shown in this work on the Florida crayfishes, I wish to express my sincerest thanks.

Especial thanks are due Dr. Waldo L. Schmitt of the United States National Museum and Dr. Fenner A. Chace of the Museum of Comparative Zo-

Map 1.—County Map of Florida showing Areas in which Field Work has been done.

ology who have kindly made comparisons of my specimens with typical material and through whom loans were secured from their respective museums. I wish also to express my gratitude to Dr. E. P. Creaser formerly of the Museum of Zoology, University of Michigan, for the identification of specimens when this work was first begun.

To Professors T. H. Hubbell, C. F. Byers, H. B. Sherman, and M. D. Cody I wish to express my appreciation for many valuable suggestions and criticisms throughout the course of the present study. Thanks are also due Mr. G. Robert Lunz of the Charleston Museum and Mr. K. E. Goellner of the Museum of Zoology, University of Michigan, for the loan of specimens.

For their contribution of specimens and for companionship on collecting trips I am indebted to Dr. H. K. Wallace, Dr. A. F. Carr, Dr. Lewis Berner, Dr. Frank N. Young, Dr. A. M. Laessle, Messrs. Lewis J. Marchand, W. M. McLane, Joel M. Martin, John D. Kilby, Coleman J. Goin, James J. Friauf, R. E. Bellamy, Glendy Sadler, G. W. Van Hyning, R. W. Williams, Ray Boles, Lee Stanton, and many others.

Finally I wish to acknowledge my great indebtedness to my parents, Mr. and Mrs. H. H. Hobbs, who have made this study possible through financial aid and by providing transportation on numerous collecting trips, and to my wife, Georgia Blount Hobbs, who has checked many of the detailed compilations necessary in the preparation of the manuscript.

Material Studied

By far the greatest amount of the material on which this paper is based is contained in my own collection, nearly all of which has been collected by myself and my colleagues of the University of Florida. This collection comprises some ten to twelve thousand Florida specimens.

I have also examined an additional three to five hundred specimens in the collections of the various museums. These included all of the extant types of the species that have been described from Florida by other workers.

In 1935 and again in 1937 I visited the United States National Museum in order to study the type specimens of all of the species of the subfamily Cambarinae contained in that collection. Further typical material was examined at the Museum of Comparative Zoology and at the Philadelphia Academy of Sciences during February 1937. While at these museums I took the opportunity of examining many other specimens which had been collected in Florida, Georgia, and Alabama.

The United States National Museum, the University of Michigan Museum of Zoology, and the Museum of Comparative Zoology have kindly lent me numerous specimens including some Florida materials. These have been extremely useful in making comparisons and in giving me additional locality records for the state.

Unless otherwise specified all of the specimens mentioned in this paper are in my personal collection at the University of Florida.[1]

In the beginning I had planned to do intensive collecting throughout the state of Florida; however, it was soon apparent that some regions required much more detailed study than others. The crayfish fauna of southern Florida is restricted to two widely distributed species, but the northern and eastern portions of the state have no fewer than 40 species or subspecies, and so required intensive collecting in order to decipher relationships and ascertain the limits of the ranges involved.

The accompanying map shows that although collections have been made in all of the 67 counties of Florida, western and northern Florida have received a much greater emphasis than the southern two-thirds of the peninsula. In addition collections have been made in the following counties of Alabama and Georgia: ALABAMA—Baldwin, Butler, Conecuh, Dale, Elmore, Escambia, Houston, Lee, Lowndes, Mobile, Montgomery, and Russell; GEORGIA—Appling, Baker, Ben Hill, Brooks, Bryan, Bulloch, Burke, Camden, Clarke, Clinch, Colquitt, Cook, Decatur, Dooly, Dougherty, Early, Echols, Effingham, Emanuel, Floyd, Glynn, Grady, Gray, Greene, Habersham, Houston, Jefferson, Jenkins, Jones, Lanier, Liberty, Long, Lowndes, McDuffie, McIntosh, Mitchell, Murray, Oconee, Pierce, Rabun, Screven, Sumter, Tattnall, Thomas, Toombs, Ware, Washington, Wayne, and Wilkes.

METHODS OF COLLECTING

One of the best collecting implements is a fine meshed minnow seine. In open water, even in small ditches or puddles where vegetation is not too dense, thorough seining is quickly completed and, if care is taken to drag the bottom, is highly efficient. Seines also proved successful in those underground pools in which the bottom was not strewn with rocks.

Another very useful device is the "D"-ring or semicircular dip net. This consists of a semicircular frame, twelve inches in diameter, made of one-quarter inch spring-steel rod with the ferrule in the middle of the rounded side, and fitted with a five foot hickory handle. A variety of bags have been used with this net, all equipped with a strong canvas rim. A bag of one-quarter or one-half inch netting permits the swiftest movement through the water and is particularly useful in subterranean water where the crayfishes are likely to be very wary and require rapid manipulation of the net. Where *C. schmitti* or other very small species are sought for it is necessary to use a much smaller mesh, and a bag of strong scrim with about one-sixteenth inch mesh is excellent. A useful net for general work has a short canvas bag with a bottom of strong wire screen (one-sixteenth inch mesh).

---

[1] The holotype and allotype of all of the species I have described or am describing in this paper are deposited in the United States National Museum. Paratypes of several of them have been deposited in the Museum of Comparative Zoology, University of Michigan Museum of Zoology, the Philadelphia Academy of Sciences, and the Carnegie Museum.

An ordinary garden rake is often very effective in roadside ditches and in streams thickly grown with vegetation. Masses and mats of vegetation when torn loose and brought out on the banks generally contain many of the crayfishes that live in these habitats. In many of the lakes and ponds of Florida crayfishes are abundant in the roots of the floating water hyacinths. By rapidly rolling the hyacinth mats or lifting the tangled mass onto the bank, many crayfishes may be captured as they crawl back toward the water. Some of the more sluggish species may readily be captured without either nets or seines, and in swift water where stones make a dip net cumbersome, it is often best to turn the stones and capture the crayfish by hand.

Probably the most satisfactory of all collecting methods is to hunt at night with the aid of a headlight. Then many of the crayfishes have left their daytime hiding places and are to be found venturing into relatively open situations. When the bright beam of the light strikes the eyes of a crayfish two brilliant red spots are reflected, and at fairly close range the entire animal is clearly visible.

A large number of the specimens in my collection were dug from burrows, and many species can rarely be collected in any other way. Since the size of the burrow, and especially the size of the chimney pellets, gives some indication of the size of the crayfish that made the burrow, it is possible to concentrate on mature specimens and largely avoid laborious delving for immatures.

After digging out a number of crayfishes one learns the habits of the various species and can often identify them before he begins to dig, and once the burrowing habits of a species are learned, the task of digging becomes less arduous. For those which construct simple burrows a shovel may be used to dig almost straight down to the bottom of the excavation, and little work with the hands is necessary. When the mouth of the burrow opens below the water table it is almost impossible to secure the specimen unless one keeps his hand in the burrow to prevent the escape of the crayfish. The digging out of all primary burrowers is more of a task. Since the burrows are complex and ramify in any direction it is necessary to keep one hand in the burrow so that none of the tunnels will be lost. Collecting is thus much easier if two persons work together. When one must work alone a garden trowel may be used to loosen the soil around the burrow so that the hand in the burrow may proceed downward until the crayfish is cornered. In some instances, where the burrows are decidedly more complex, one must dissect the entire structure; more often the crayfish seeks the deepest portion, but occasionally it retreats to one of the side passages. The specimen should be grasped by the carapace, but often it is the crayfish which seizes the collector's finger. In the latter case the crayfish can often be drawn to the surface or far enough up the burrow to be readily caught. For certain of the primary burrowing species the extraction of the crayfish from its burrow is much simpler. After the water table is reached, the water in the burrow is stirred vigorously and then left undisturbed for a short time; soon the antennae of the crayfish will be seen whipping to and fro at the surface, and if

```
                3-1839-2a, 2b                    March 18, 1939
                Tallahassee                      Leon County, Florida

                    12 miles west of Tallahassee on St. Hy. 19. Crypto-
                labis Ravine - a deep ravine leading into the Ochlocknee
                River. Water issues from boggy seepage area and flows
                into a small clear stream with a coarse sandy bottom.
          a1    P. spiculifer was taken from the creek, under logs and
                debris while C. floridanus was taken with considerable
                difficulty from burrows of the stream beds. The burrows
                are extensive, and the thick mat of roots makes digging
                a task. (Other pertinent data are included here).
                P. spiculifer (LeConte)
                C. floridanus (ms. name)
                              Rogers, Young, Berner, and Hobbs
```

a2: Procambarus spiculifer — Detrm. by Hobbs 2-20-1940; Number 3-1839-2a; State Florida; County Leon; Locality 12 miles west of Tallahassee on St. Hy. 19. Cryptolabis Ravine; 1 ♂ I, 14 ♂ II, 10 ♀♀, 1 ♂ imm., 2 ♀ imm., 1 ♀ eggs, — ♀ young; Crayfish Associates: Cambarus floridanus; Collector Hobbs; Collection Hobbs

a3: Cambarus floridanus — Detrm. by Hobbs 2-20-1940; Number 3-1839-2b; State Florida; County Leon; Locality 12 miles west of Tallahassee on St. Hy. 19. Cryptolabis Ravine; 1 ♂ I, 2 ♂ II, 1 ♀, — ♂ imm., 2 ♀ imm., — ♀ eggs, — ♀ young; Crayfish Associates: Procambarus spiculifer; Collector Hobbs; Collection Hobbs

Text Figure a.—Specimens of Field and Species Catalogue Cards.

one makes a quick grab the crayfish may be easily captured. If the first grab is unsuccessful the crayfish will return to the surface in a short time.

The cavernicolous species, of course, present special collecting problems due to limited and varied access to their subterranean habitats. This is emphasized in the case of *Troglocambarus maclanei*, which has the very unusual and apparently specific habit of clinging to the ceilings of wholly submerged portions of underground waterways. These various collecting methods are described under the individual discussions of the cavernicolous species.

Although trapping has not been sufficiently tested, the few attempts that I have made proved very unsatisfactory. An ordinary minnow trap made of screen wire with a funnel turned inward at one end worked fairly well in a cave, but I have had little success in trapping any of the surface species.

Field Notes

Detailed field notes with precise data as to locality and habitat are particularly pertinent in Florida. The abrupt changes of soil types and ground water often make an apparent discontinuity in the range of a certain species very hard to verify, and it is often difficult or impossible to discover without detailed habitat data just what factors in the environment are correlated with a species' ecological distribution.

The kind of data recorded in my field catalogue which is kept on 3" x 5" library cards is illustrated in figure a-1. The catalogue number "3-1839-2a, 2b" is written in the upper left-hand corner; the figure "3" represents the month (March); the next (one or) two digits "18" denote the day of the month; the next two indicate the year (1939). The figure "2", following the second dash indicates that this was the second collection made on March 18, 1939. (The "a" and "b" following the "2" show that two different species were collected from this station). Text figures a-2 and a-3 are the references to this field catalogue entry in the species catalogue.

Methods of Preservation and Study

After trying several methods of killing and preserving crayfishes the following has been adopted as the most satisfactory standard process: as soon as the specimens are caught they are dropped into 80% ethyl alcohol; after several days all except a quarter of an inch (left to avoid losing any branchiobdellid worms) is poured off and enough fresh alcohol of the same percentage is added to well cover the specimens. This, if the container is tightly sealed, will last for several years. All of my specimens except large series are kept in eight ounce, clamp-top, glass jars. This method of preservation has the disadvantage that the color of the animal disappears almost immediately. A four percent solution of formaldehyde is less rapid in destruction of color, and in some instances I have had specimens preserved in formaldehyde in which the color pattern, but not the color, was evident even a year after preservation, but formaldehyde makes the specimens extremely brittle, so that a great deal of breakage is inevitable. The 80% alcohol, on the other hand, leaves all of the joints pliable so that the specimen may be handled with little danger of being broken, and to work with such specimens, of course, is much more pleasant. For studies of the first pleopod of the male or annulus ventralis of the female it is best to remove the part in question, dehydrate with alcohol, clear it with xylene, and after allowing it to dry, mount it on an insect pin with a numbered label that permits one to unquestionably associate the pinned part with the proper specimen. I have never found any shrinkage in such cleaned and pinned mounts.

Previous Work on the Crayfishes of Florida

The earliest published record of a crayfish taken from Florida was that of R. Gibbes in 1850. Since this record was for *C. affinis* the determination was evidently an error, and what species was actually seen is left for conjecture.[2]

Hagen (1870) described *C. fallax* and recorded *C. lecontei*. The specimens on which the latter determination was made were collected at Pensacola, and although I have not seen them, it is probable that they are either *versutus* or *spiculifer*, but hardly *lecontei*.

Faxon (1884) recorded *alleni, clarkii, fallax,* and *versutus* from Florida but stated "three young specimens from Pensacola, Fla. (M. C. Z., No. 249), are also placed here [in *C. lecontei*] by Hagen, but the justice of the determination seems doubtful. They do not agree with the types from Mobile." Then (1885a) in the *Revision of the Astacidae* he gave additional locality records for *fallax* and *clarkii* and stated that "a species of the *C. bartonii* group also inhabits Florida." In 1890 he added an additional locality for *alleni*, gave a brief description of its female, and described *C. evermanni* from Flomaton which is on the Alabama-Florida line. From the same locality he recorded *barbatus* (see p. 40).

Lönnberg (1894a) made a preliminary report on his collecting *fallax* and *alleni* from Florida and on the discovery of *C. acherontis*. In his second publication (1894b) he cited additional localities for *fallax* and *alleni* and described *C. acherontis*, the first of the cavernicolous species known from Florida.

Faxon (1898) recorded additional localities for *fallax, alleni,* and another species that he took to be Lönnberg's *acherontis*.[3]

Harris (1903) in his *An Ecological Catalogue of the Crayfishes Belonging to the Genus Cambarus* listed five species, *fallax, clarkii, versutus, alleni,* and *acherontis*, from Florida.

---

[2]Hagen (1870: 101) "Lewis R. Gibbes quotes *C. affinis* from Florida, but his determinations are not at all trustworthy."

[3]This is not *acherontis* but *lucifugus lucifugus* (Hobbs 1940: 389).

Ortmann (1905a) stated that six species are known from Florida and listed *evermanni* and *barbatus* as occurring in the western part of the state, as well as in Alabama and Mississippi, although he cited no Florida records for these two species.

Faxon (1914) gave additional localities for *fallax* and *alleni* and described the peninsula specimens heretofore referred to *clarkii* as *clarkii paeninsulanus*. The additional locality cited for *acherontis*[4] (Eustis, Lake County, Florida) is again based on an erroneous determination.

Subsequent to 1914 no further references to Florida crayfishes were published until I described *Cambarus rogersi* in 1938. Thus in 1932 when my own work on Florida crayfishes began ten species had been recorded for Florida: *affinis, fallax, lecontei, alleni, clarkii, versutus, acherontis, barbatus, evermanni*, and *clarkii paeninsulanus*. Of these, one, *affinis*, must be deleted; two others, *clarkii* and *barbatus*, should in all probability be now referred to closely allied but distinct species,[5] and a fourth, described under the trinomial *Cambarus clarkii paeninsulanus*, is distinct from *clarkii* (*Procambarus paeninsulanus*, Hobbs and Marchand, in press).

Since 1938 my own descriptions of Florida specimens have added 13 species and subspecies to the published list of Florida crayfishes: 1940—*Cambarus pallidus, C. lucifugus lucifugus, C. lucifugus alachua, C. hubbelli, C. kilbyi, C. rathbunae*, and *C. pictus*. 1941—*C. cryptodytes, C. floridanus*, and *C. byersi*. I also have two papers which are now in press describing a new genus, *Troglocambarus*, and three new species: *T. maclanei, Procambarus pycnogonopodus*, and *P. leonensis*.

FACTORS AFFECTING THE DISTRIBUTION OF CRAYFISHES IN FLORIDA

Florida as a Faunal Region

Cooke and Mossom (1929), Cooke (1925, 1939), Leverett (1931), and Stubbs and Hubbell as quoted by Carr (1940) have discussed or summarized what is known or conjectured about the geological history of Florida, and I have drawn freely from them in my discussion of the possible migration routes of the several invading crayfish stocks. A great difficulty lies in the dearth of conclusions or of even a preponderance of evidence for the sequence of the minor advances, retreats, and embayments of the gently rising and sinking Florida coasts. That such events took place is evident enough, but the time relationship between the coastal terraces and other minor changes is not clear; nor is it definitely known whether the early Pleistocene submergences involved all of the peninsula of Florida or whether certain islands in the penin-

---

[4] These specimens are near *lucifugus lucifugus* (Hobbs 1940: 393).

[5] The specimens from the Pensacola region referred to *clarkii* are herein described as a new species, *P. okaloosae*. The Florida *barbatus* mentioned by Ortmann (1905a) is probably *escambiensis*. Specimens determined *lecontei*, as stated above, were probably confused with *spiculifer* or *versutus*.

sula region have had a continuous terrestrial existence since sometime in the Pliocene. This much at least is certain, that Florida, as well as the adjacent parts of Georgia and Alabama, is geologically recent and ecologically young, and that the migration of crayfish into and within the state must have largely taken place subsequent to early Pleistocene and is still in progress.

It seems best to qualify the recentness of migration with a "largely" because of the marked structural changes and distribution shown by the four peninsular cavernicolous forms. St. John (1936), Hubbell (1939), and Carr (1940) have presented good biogeographical evidence for the persistence of an Ocala Island since late Pliocene, in spite of strong geological evidence that it must have been submerged. Certainly the existence of such an island and the survival there of an earlier wave of migration from which the cavernicolous forms, and probably *Procambarus pictus* as well, were derived would give a far more comfortable margin of time for the development of the several cavernicolous species that now occupy the site or periphery of the Ocala Island. Indeed it is not unthinkable that adaptation to subterranean waters which reached the "Island" by way of strata from the Piedmont fitted the crayfish to utilize this freshwater refuge during a Pleistocene submergence, if Cooke's contention is to be accepted.

On the other hand, cavernicolous adaptations may not require more time than has been available since the end of the early Pleistocene. Certainly limestone solution in Florida is rapid and extensive and must have begun soon after Florida appeared above the sea. In the light of the marked tendency to develop spelean adaptations shown by the crayfish as a group, and in the light of recent indications from genetics and Kinsey's conclusions (1929, 1936) on evolution in *Cynips* it may well be that there has been sufficient time since Pleistocene to provide for the invasion and adaptations of all the Florida fauna.

Here in the southeastern coastal plain topographic relief, although slight, is accentuated by an extensive variation in soil types; upland soils are irregularly interspaced with poorly drained flatwoods; extensive areas of swamps and marsh are broken or wholly segregated by well drained sands; and the remarkable development of subterranean drainage with its development of caverns, huge springs, sink holes, and thousands of solution ponds and lakes has also in many regions accentuated the xeric condition of many wide extents of surface soil.

The chief extensive areas of well drained soils that appear to have been the most generally important barriers to migration are shown on Map 2. Other local barriers as well as the chief highways for dispersal are best discussed under the particular species or groups concerned.

DERIVATION OF THE FAUNA

The crayfish fauna of Florida comprises five genera which have had a relatively long and complex history. According to Ortmann the Cambarinae (his genus *Cambarus*) "originated in Mexico, and immigrated, probably at the beginning of the Tertiary, into the southwestern and southern United States, originally occupying only the southwestern Cretaceous plain, the Ozark Mountains, and the southern extremity of the Appalachian System."

He cites the following centers of origin for the various genera: *Procambarus* (Ortmann's subgenera *Cambarus* and *Procambarus*) originated in Mexico and following the above mentioned path reached the foot of the Appalachian Mountains in the lowlands of Alabama and Georgia where a "secondary center" was developed. "Here the more advanced forms of this subgenus took their origin, and spread all over the Atlantic and Gulf coast plain, and further up the Mississippi valley" (Ortmann 1905a: 124).

Cambarellus (Ortmann's subgenus *Cambarellus*) "originated in the southern United States and immigrated into Mexico, first into the central plateau, then into the Pacific slope" (Ortmann 1906a: 24).

Orconectes (Ortmann's subgenus *Faxonius*) "developed in the central basin of the three great rivers, spreading over almost all of the Mississippi drainage, and crossing over into the Hudson Bay, Great Lakes, and even into the Atlantic drainages, probably by the aid of shifting divides" (Ortmann 1905a: 24).

Cambarus (Ortmann's subgenus *Bartonius*) came into existence "in the mountainous region of the southern Appalachians, probably including also the Ozark region, and from here it spread chiefly over the Appalachian chain in a northeasterly direction as far as New Brunswick" (Ortmann 1905a: 124).

The fifth genus, *Troglocambarus*, probably had its origin in the cavernicolous waters of the Florida peninsula, and is possibly of much more recent origin than the others.

Some representatives of *Procambarus* or its progenitors migrating into the southeast pushed southward and then followed the Gulf coast as the sea retreated, so that when the elevation of Florida left vast stretches of poorly drained lowlands, traversed by numerous streams, a broad highway was opened to them. Thus the *Procambarids* reached Florida from the north along several paths. These paths are pointed out below in the section devoted to Principal Migration Routes.

The genus *Cambarellus* is represented in Florida by only one species, which occupies a large range in the panhandle and south and east as far as the drainage of the Suwannee River. This stock undoubtedly came into Florida from the west, making its way along streams and through the coastal flatwoods.

Likewise only one species of the genus *Orconectes* occurs in Florida, and the available data indicate that this species is a migrant from the west, which has taken a path similar to that followed by *Cambarellus*. However, *Orconectes* took a more northern route and penetrated into the state only in the West Florida Lime-Sink or Cypress Pond Region (Harper 1914: 201).

The genus *Cambarus* reached Florida along at least three different routes: one of them from the west along the coastal region, the others from the north, one in the region of the Escambia River, and the other along the Apalachicola River. None of the members of this genus are found east of the Ochlocknee River in Gadsden County.

Several authors, notably, Hubbell (1940), Rogers (1933), Carr (1940), and Kurz (1933), have been impressed by the fauna and flora of the remarkable ravines along the east bank of the Apalachicola River in Liberty and Gadsden counties. These authors have pointed out that several animals and plants which are common in the Appalachian and Piedmont regions reach the southern limits of their ranges in this area, and in addition that several forms, endemic in the ravines, have their closest affinities with northern groups. At least two plants, *Tumion taxifolium* and *Taxus floridana*, are probably Tertiary relicts, while the katydid *Hubbellia marginifera* (Walker) has relatives only in Europe, the Near East, and Madeira. Carr (1940: 4) presents a very plausible explanation of why the Apalachicola River supports such a large assemblage of animals and plants having their closest affinities with Appalachian and Piedmont species.

Endemism

Twenty-five of the 42 Floridian Cambarinae are, as far as known, endemic to the state. After thorough collecting is done in southern Alabama this probably will be reduced to not more than 17. Of these 17 six are cavernicolous forms.

The following are now recorded only from Florida but probably occur in southern Alabama: *P. escambiensis, P. rathbunae, P. hubbelli, P. shermani, P. okaloosae, P. pycnogonopodus, P. bivittatus, C. species incertis*. The following 17 are probably confined to Florida: *P. apalachicolae, P. econfinae, P. latipleurum, P. kilbyi, P. alleni, P. geodytes, P. rogersi rogersi, P. rogersi campestris, P. rogersi ochlocknensis, P. acherontis, P. youngi, P. pictus, P. lucifugus lucifugus, P. lucifugus alachua, P. pallidus, C. cryptodytes, T. maclanei*.

The Floridian cavernicoles are representatives of at least four independent advances into the subterranean waters. Three of these were probably correlated with the early insular development in central Florida. It seems most probable that the stock which gave rise to *Procambarus acherontis* reached the Pliocene islands at an early time but was destroyed by subsequent emergence leaving *acherontis* as a relict in the subterranean waters of the area. The cavernicolous species of the *pictus* group *(P. lucifugus lucifugus, P. lucifugus alachua, P. pallidus)* and *Troglocambarus maclanei* (the latter probably of earlier origin) have evidently been derived from a stock of stream dwellers which occupied the southeastern coastal plain and migrated into the peninsula when the Suwannee Straits were closed. The only spelean species of *Cambarus (C. cryptodytes)* found in Florida is probably a derivative of an earlier Appalachian and Piedmont stock which moved southward along the Apalachicola River.

At least two surface dwelling species of *Procambarus* are probably relicts of a once widely distributed stock. These are *P. alleni*, occupying an extensive area in the southern part of the peninsula, and *P. pictus*, confined to a few small tributaries of the St. Johns River in Clay County. The latter is particularly interesting in that its closest affinities are found among the peninsular cavernicoles.

The other endemics are derived from three stocks which have apparently migrated into the region in more recent times. One of these, the *barbatus* stock, (illustrated on Map 3) reaching Florida along several migration paths, has given rise to four species in the panhandle: *P. apalachicolae*, *P. econfinae*, *P. latipleurum*, and *P. kilbyi*. The *pictus* stock, which probably had its origin in the southeastern part of Georgia, migrated southwestward along the Ochlocknee River system into the coastal flatwoods, crossed the Apalachicola and gave rise to *P. youngi*, which appears to be extremely localized in a few small streams around Weewahitchka. The *advena* stock following the same route produced *P. pygmaeus* and the three subspecies of *P. rogersi*.

Principal Migration Routes

I believe there is sufficient data to postulate the principal migration routes by which the crayfishes have reached Florida and I have indicated the species, or their forerunners, which appear to have utilized each route.

I. Early Migrations During Former Bridgings of the Suwannee Straits

It seems probable that the Straits were reopened at least twice after insular Florida was first connected with the mainland. The first openings of the Straits with the probable subsequent submergence of the whole peninsula would provide for the origin of the cavernicolous *P. acherontis*. The second bridging of the Straits provided for the entrance of the stocks that gave rise to *T. maclanei*, *P. pictus*, and the subterranean species: *P. lucifugus lucifugus* and *P. lucifugus alachua*. Perhaps the *alleni* stock made its way into the peninsula at the same time. These stocks were thus left to occupy the Florida island after the third opening of the Straits barred it for a time to further immigration.

II. Migrations Since the Last Closing of the Suwannee Straits

1. Migration Southward in the Flatwoods Region along Trail Ridge. Four Florida species have apparently entered or been derived from stocks that entered the state along this path. These are the endemic *P. geodytes*, and *P. seminolae*, *P. pubischelae*, and *P. advena*.

2. Immigrations Via River Systems and Adjacent Flatwoods.
    (a) Perdido River: *P. escambiensis*, *P. evermanni* (?), *P. spiculifer*, *P. versutus*, and *C. byersi*.

(b) Escambia River: *P. bivittatus, P. shermani, P. blandingii acutus, P. evermanni, P. okaloosae, P. spiculifer, P. versutus,* and *C. species incertis.*

(c) Apalachicola and Chipola rivers: *C. latimanus, C. floridanus, P. apalachicolae, P. latipleurum, P. spiculifer,* and *P. versutus.*

(d) Ochlocknee River which acts as a connecting link between the lowland of southeast Georgia and the Apalachicola flatwoods: *P. youngi, P. pygmaeus, P. rogersi rogersi, P. rogersi ochlocknensis, P. rogersi campestris,* and *P. kilbyi.*

(e) Choctawhatchee River and the Marianna Lowlands: *P. hubbelli* and *O. clypeata.*

Several of the stream inhabitants are so widely disseminated throughout the streams of the state that I cannot reconstruct their paths of entry into the Florida waters with any feeling of certainty, and indeed a number of these stream inhabitants have undoubtedly entered the state by several paths.

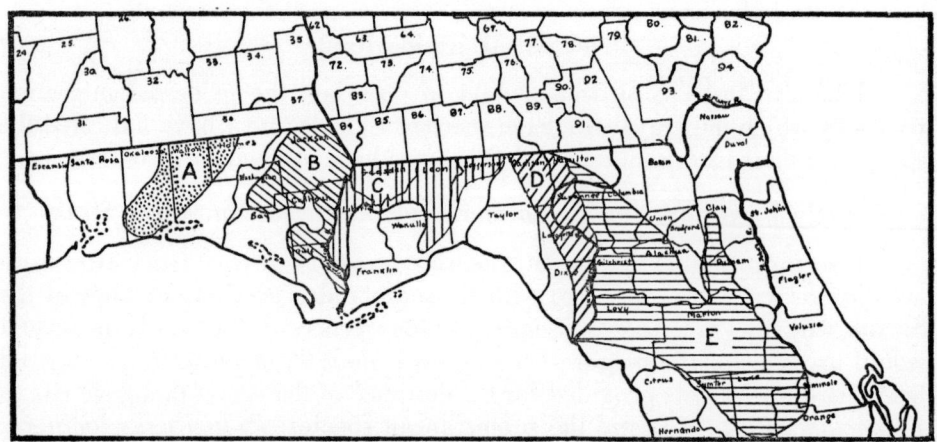

Map 2.—Continuous North-South Extents of Well Drained Soils that appear to act as Barriers to Crayfish Migration.

CONTINUOUS NORTH-SOUTH EXTENTS OF WELL-DRAINED SOILS THAT APPEAR TO ACT AS BARRIERS TO CRAYFISH MIGRATION

(Explanation to Text Map 2)

A. (Norfolk-Greenville, and Norfolk-Orangeburg Soil Areas)
Western barrier to: *P. paeninsulanus, P. hubbelli, P. pycnogonopodus, P. apalachicolae.*
Eastern barrier to: *P. byersi, P. okaloosae, P. evermanni* (part).

B. (Norfolk-Greenville, and Norfolk-Orangeburg Soil Areas)
   Western or northern barrier to: *P. rogersi rogersi, P. latipleurum,*
                                    P. kilbyi (part).
   Eastern barrier to: *P. hubbelli* (part), *P. latipleurum.*

C. (Norfolk-Greenville, Blanton-Norfolk, and Norfolk-Orangeburg Soil Areas, reinforced at the south by the broad mouth of the Apalachicola)
   Western barrier to: *P. rogersi ochlocknensis, P. leonensis.*
   Eastern barrier to: *P. seminolae, P. pycnogonopodus, P. youngi.*

D. (Blanton-Norfolk Soil Areas)
   Western barrier to: *P. fallax, P. alleni* (?).
   Eastern barrier to: *P. leonensis.*

E. (Blanton-Norfolk, and Fellowship-Gainesville Soil Areas)
   Southern barrier to: *P. pubischelae, P. seminolae, P. pygmaeus*
                         (part), *P. advena.*
   Northern barrier to: *P. alleni.*
   Eastern barrier to: *C. schmitti, P. spiculifer* (part).
   Western barrier to: *P. pictus.*

Species That Traverse Such Barriers or Occur on Either Side of Them

A number of species are found on both sides or within the limits of the barriers in question. Their means of traversing them are indicated by: N—the present range appears to extend north of the northern limits of the barrier; S—the range has been extended into coastal flatwoods south of the southern extent of the barrier; ?—means of traversing barrier definitely unknown.

Barrier A—
   P. rathbunae—? (N)
   P. spiculifer—N
   P. versutus—? (N)
   C. schmitti—?
   C. diogenes—N

Barrier B—
   P. paeninsulanus—?
   P. pycnogonopodus—S
   P. spiculifer—N
   P. versutus—?
   C. schmitti—?
   O. clypeata—N
   C. diogenes—N

Barrier C—
   P. kilbyi—S
   P. paeninsulanus—?
   P. pygmaeus—S
   P. rogersi (complex)—S
   P. spiculifer—N
   C. schmitti—?

Barrier D—
   P. kilbyi—S
   P. paeninsulanus—?
   P. spiculifer—N
   C. schmitti—?

Barrier E—
   P. kilbyi—S
   P. fallax—?
   P. paeninsulanus—?
   P. spiculifer (part)—N

Ecology

In general, it may be said that crayfishes are found in any body of freshwater and in almost any poorly drained soil. Small local areas of apparently suitable soils are occasionally found to be unoccupied, but these areas will usually prove to be completely surrounded by well drained soils and thus isolated from any natural stocking.

In attempting to classify and describe the various types of situations inhabited by the several species of crayfishes, one is confronted with the difficulty that the actual distinctions between the situations occupied by different species are complex, and often not at all easy to discern. The differences and likenesses beween situations that are most apparent to the student are by no means always the distinctions that appear to be most important to the crayfish. Nevertheless the desirability of attempting to describe and classify habitats in terms that can be recognized by the field naturalist is very real. In the discussion that follows, the ecological subdivisions are frequently more detailed than the actual habitat discrimination shown by the crayfishes would warrant and result in many species being listed in two or more distinct situations. So detailed a subdivision, however, at least facilitates the generalizations that are necessary if one is to attempt a summary of the habitat correlations of the whole crayfish fauna.

Most of the difficulty in defining a clear-cut ecological distribution of the whole fauna is due to two circumstances: that Florida comprises several regions that are actually quite diverse in topography, soil, and drainage, and that for so many species the geographic range does not coincide with any major ecological subdivision of the state.

Aquatic Habitats

Lotic Situations

Several large rivers, such as the Apalachicola and Choctawhatchee, which cross the panhandle of Florida have their origins in the agricultural sections of Alabama and Georgia and carry heavy loads of silt and clay which give them the characteristic red color of streams that rise in the Piedmont. For the most part the beds and margins of these streams are nearly barren of aquatic plants, but extensive hardwood swamps are developed in their flood plains. Such lower streams undoubtedly have a larger crayfish fauna than is indicated by the number of specimens recorded from them, but I believe that they do not support so abundant a fauna as their smaller tributaries. However, the difficulties of collecting have prevented as thorough a survey of the large lower streams as has been possible in the smaller water courses.

Several of the smaller rivers, for example, the Chipola and Santa Fe, are partly spring-fed and flow through lime rock outcrops along part of their courses. Only after heavy rains do these streams become turbid or take on the dark red or brown color of swamp water. In places they flow over limestone outcrops and produce shoals and riffles, and in many of these such aquatic vegetation as *Vallisneria, Sagittaria, Philotria, Riccia,* and *Chara* is abundant. A more complex description of this type of stream is given on page 123. Although the number of species of crayfishes is small, the population is dense, especially in the riffles and vegetation zones.

The Hillsborough, Manatee, and Withlacoochee rivers represent another class of the larger streams. These rivers arise in the swampy lake regions of the interior of the peninsula, and for the most part are devoid of aquatic plants. In some of the broader areas, however, *Nymphaea, Piaropus,* and other aquatics associated with lenitic situations are abundant, and in shallow reaches some of these streams gain momentum and form riffles which support as luxuriant a flora as that of the spring-fed rivers of the panhandle and northern Florida. The crayfish population in the shallow and broad regions of these streams is extremely large.

The flatwoods streams of the panhandle and northern Florida are numerous and in general have many characters in common. Most of them are small and relatively shallow, supporting dense growths of *Myriophyllum, Potamogeton, Juncus, Pontederia, Sagittaria,* and filamentous algae. Most of these streams are sluggish, with dark water indicating an abundance of organic acids. Many are semi-permanent and flow only part of the year. The crayfish population reaches a maximum in this type of stream. If the water ceases to flow and the stream begins to dry up the crayfishes burrow or live in the deeper pools where the water is more permanent.

Scattered throughout the state, particularly in the rolling uplands, many small, clear, sand bottomed streams occur. This type of stream is abundant along the north shore of the Choctawhatchee Bay, in the rolling country east of Crestview, Okaloosa County, and in the Central Highlands. These streams originate in seepage areas and small springs and flow through sandy and clay beds over most of their course. Aquatic vegetation consisting of *Orontium, Nymphaea,* and *Potamogeton* is sparsely scattered along the margins. If one collects in these streams during the day when the crayfishes have retreated into their burrows in the banks or remain concealed in the roots of the marginal semiaquatic vegetation, he is left with the impression that the crayfish fauna is poor; however, at night the animals appear in large numbers.

Numerous large springs are present throughout the state. Most of them flow into large basins which support a rich aquatic flora that includes *Nymphaea, Riccia, Cabomba, Potamogeton, Isnardia, Cardamine, Myriophyllum, Utricularia, Chara,* and filamentous algae. Many of these springs (viz., Blue Springs, Jackson County, Wakulla Springs, Wakulla County, Silver Springs, Marion County, etc.) form the origin of long "spring runs" where aquatic plants, especially *Vallisneria, Naias, Sagittaria,* and *Ceratophyllum* are abundant. Crayfish are numerous in the vegetation both in the springs and in the runs. A number of the large springs in the peninsula flow from or through sulfur- and salt-bearing strata and deposit dense, cream-colored sludges and precipitates on the vegetation in them. Even here *Procambarus fallax* is often common.

In the deep ravines along the east side of the Apalachicola River in Liberty County, small streams that have their origins in seepage areas and small springs flow over barren sand and clay bottoms and are often dammed by fallen limbs and leaf drifts. These streams are all well shaded and cool and support a surprisingly rich crayfish fauna.

Lenitic Situations

Temporary ponds are very common in all parts of the state, both in the rolling uplands and in the flatwoods areas. In both regions the ponds vary widely as to the time and regularity of becoming dry. Some are dry except during the rainy season, and seldom have a crayfish population; others which become dry for only a short period during the year, support a large population. When these ponds do become dry the crayfishes construct burrows. Some of these temporary ponds are within dense swamps and are barren of aquatic plants, while in the more open ponds *Sagittaria, Panicum, Juncus, Utricularia,* and filamentous algae are often abundant.

Permanent but fluctuating ponds are also common in most regions of the state. Such ponds rise and fall markedly with the rainy and dry seasons and usually have an abundance of rooted and floating vegetation. Some of the commoner plants are *Nymphaea, Castalia, Piaropus, Pontederia,* and *Ceratophyllum*. These ponds usually support a very large crayfish population.

Deep, well-like sink-hole ponds are common in certain regions of Florida and many are inhabited by crayfish, but ponds of this type do not have nearly so abundant a crayfish fauna as the shallower ponds which support a more luxuriant growth of hydrophytes.

In the larger lenitic situations the crayfish population appears very definitely to be correlated with the kind and amount of rooted aquatic vegetation. Although from the standpoint of the limnologist the lakes of Florida fall into several rather distinct types, ranging from the deeper, clear lakes of the sand hill region to the shallow and usually acid and discolored waters of the flatwoods lakes, for the crayfish it appears to be only the character of the shoreline that is important.

The deep and often extensive Nymphaea-Nelumbo-Utricularia marshes, that are locally known as "Prairies", are common in many parts of Florida. They usually have a depth of about six feet or less, and almost the entire area is crowded with a dense growth of both emergent and submergent vegetation, with *Nymphaea, Piaropus, Nelumbo, Typha, Juncus,* and grasses especially common. Here a few species of crayfishes are often extremely abundant.

Roadside ditches and canals in which water stands most of the year also develop a rich growth of submergent, emergent, and floating aquatic plants. Commonly found in these plant associations are *Myriophyllum, Ceratophyl-*

lum, Potamogeton, Utricularia, Persicaria, Ludwigia, Pontederia, Sagittaria, Nymphaea, Castalia, Piaropus, Globifera, and grasses. These ditches which form a very typical and characteristic part of much of the Florida landscape are one of the most productive of all crayfish habitats.

In the everglades freshwater marshes are extensively developed with saw grass as the characteristic dominant. Although there are numerous pot holes which dot the rock ridge bounding the everglades along the east coast, the most extensive bodies of fresh water are the abandoned rock pits and the extensive canals. In these *P. alleni* often occurs in large numbers.

Subterranean Situations

Much of both the peninsular and panhandle regions have extensive underground water systems, which in many cases have been exposed by the formation of sink holes and small limestone caverns. In a number of places where these cavernicolous waters are accessible various spelean species of crayfishes have been taken. These underground waters, at least throughout the peninsula, show very little variation in either temperature or pH, being around 70° and 7.1 respectively. Data on the temperature and pH of the underground waters of the panhandle is very meager but indicate that these waters are very close to those of the peninsula. One subterranean species occurs in a sulfur spring (Palm Springs) near Orlando, in Seminole County. A description of this spring is given on p. 92.

Flatwoods and Seepage Areas

Perhaps the most interesting members of the crayfish fauna of Florida are the burrowing species, which are largely confined to the flatwoods and seepage areas. Extensive areas of flatwoods occur in the coastal regions and portions of the interior. Here the water table in many places is almost at the surface, and permanently wet areas which support a large number of semi-aquatic plants are common. Conspicuous among these plants are *Sarracenia, Ericaulon, Xyris, Drosera, Pinguicula, Frimbristylis, Rynchospora,* and a large variety of grasses. The soils of these wet areas usually belong to the Plummer and Portsmouth groups. In some of these areas the ground is literally riddled with crayfish tunnels, and chimneys are a conspicuous feature of the surface. In much of the flatwoods region characterized by Leon soils the water table periodically descends much farther below the surface, but only occasionally does it drop to a depth greater than three or four feet. While the crayfish are not so abundant in Leon soils they are present in large enough numbers to be classified as common. Large colonies of many of the flatwoods species are often found burrowing in roadside ditches; particularly is this true in the vicinity of the numerous flatwoods ponds or bays.

Seepage areas along hillsides or stream slopes are, particularly in the panhandle, usually thickly populated by burrowing crayfishes. The hillside conditions are for the most part not markedly different from those found in the wet flatwoods areas, but in the seepage areas along streams the crayfishes construct a maze of intricate tunnels among the thick tangle of tree and shrub roots. The soils in these seepage areas show considerable variation, although most of them are dark and contain a high concentration of decayed organic matter.

Ecological Habits and Habitats

A—Species That Occur In Burrows

These burrowing forms show three more or less distinct degrees of burrowing habit, indicated by: P—primary burrowers (restricted to burrows); S—secondary burrowers (generally occupying burrows but wandering into open water during rainy seasons); T—tertiary burrowers (burrowing only in periods of drought or occasionally, but not necessarily, during the breeding season).

I. *Burrowing in Seepage Areas*
   P. advena—P
   P. geodytes (around sulfur springs)—P
   P. rogersi rogersi—P
   P. rogersi campestris—P
   P. rogersi ochlocknensis—P
   C. latimanus—S
   C. floridanus—P
   C. diogenes—P or S
   C. byersi—P

II. *Burrowing in Flatwoods*
   P. advena—P
   P. pygmaeus—?
   P. rogersi rogersi—P
   P. rogersi campestris—P
   P. rogersi ochlocknensis—P
   P. pubischelae—S
   P. escambiensis—S
   P. econfinae—S
   P. latipleurum—S
   P. apalachicolae—S
   P. rathbunae—S
   P. kilbyi—S
   P. hubbelli—S
   P. alleni—S
   P. seminolae—S-T
   P. paeninsulanus—T
   P. okaloosae—T
   P. fallax—T
   P. leonensis—T
   P. pycnogonopodus—T
   O. clypeata—T (?)
   C. byersi—P

III. *Burrowing in Floodplains of Large Streams*
   C. floridanus—P
   C. diogenes—P-S
   P. shermani—S
   C. species incertis—S (?)
   P. bivittatus—T
   P. blandingii acutus—T

IV. *Burrowing in Banks of Streams*
   P. hubbelli—S
   P. kilbyi—S
   P. paeninsulanus—T
   P. spiculifer—T
   P. versutus—T

V. *Burrowing in Drying Ponds*
   P. alleni—S
   P. hubbelli—S
   P. kilbyi—S
   P. latipleurum—S
   P. seminolae—S-T
   P. fallax—T
   P. leonensis—T
   P. paeninsulanus—T
   P. pycnogonopodus—T

B—Species Occasionally or Usually Found Outside of Burrows

I. *Inhabiting Ponds, Lakes, and Ditches.* L—permanent lakes or ponds; T—temporary or fluctuating ponds; D—semipermanently inundated ditches.
   P. okaloosae—D
   P. pubischelae—D
   P. pycnogonopodus—D
   P. rathbunae—D
   P. econfinae—D-T
   P. kilbyi—D-T
   P. latipleurum—D-T
   P. seminolae—D-T
   P. alleni—T-D
   P. hubbelli—T-D
   O. clypeata—T-D
   P. paeninsulanus—L-T-D
   P. fallax—L-T-D
   P. leonensis—L-T-D

II. *Inhabiting More or Less Lotic Situations.* A—rills and ravine brooks (of Apalachicola drainage); I—intermittent streams; B—brooks, creeks, and small rivers.
   C. latimanus—A
   P. kilbyi—I
   P. okaloosae—I
   O. clypeata—I
   P. blandingii acutus—B
   P. pictus—B
   C. schmitti—B
   P. spiculifer—B
   P. versutus—B
   P. bivittatus—I-B
   P. fallax—I-B
   P. leonensis—I-B
   P. paeninsulanus—I-B
   P. pycnogonopodus—I-B
   P. seminolae—I-B

III. *Inhabiting Sluggish Streams and Sloughs*
   P. kilbyi
   P. hubbelli
   P. alleni[6]
   P. fallax
   P. leonensis
   P. pycnogonopodus
   P. bivittatus
   P. paeninsulanus
   P. okaloosae
   P. evermanni
   P. seminolae
   O. clypeata

C—Species Confined to Subterranean Waters

   P. acherontis
   P. lucifugus lucifugus
   P. lucifugus alachua
   P. pallidus
   T. maclanei
   C. cryptodytes

---

[6]Found extensively in drainage canals of southern Florida.

SUMMARY OF THE GENERIC AND SUBGENERIC CHANGES
(1870-1941) (Hobbs, in press-c)

Hagen 1870	Faxon 1885	Ortmann 1905-6	Fowler 1911	Faxon 1914	Creaser 1933	Lyle 1938	Hobbs (in press-c) (Generic Names)
Group III, in part		*Procambarus*	*Procambarus*	Group I		*Girardiella*	*Procambarus*
Group III, in part	Group I	*Cambarus*	*Ortmannicus*	Group II			
Group I, in part	Group II			Group III			
			Paracambarus	*Paracambarus*	Group IV		*Paracambarus*
	Group V	*Cambarellus*	*Cambarellus*	Group V			*Cambarellus*
Group II	Group IV	*Faxonius*	*Faxonius*	Group VI	*Faxonius*, as a generic name.		*Orconectes*
Group III, in part	Group III	*Bartonius*	*Cambarus*	Group VII			*Cambarus*
							Troglocambarus

THE SUBFAMILY CAMBARINAE

All of the North American crayfishes east of the Rocky Mountains belong to the subfamily Cambarinae of the Family Astacidae. As here recognized this subfamily comprises six genera. One of these, the monotypic *Paracambarus*, is found only in Mexico; all of the other five occur in the United States east of the Rocky Mountains, and of these, two extend northward into Canada; one is found in Cuba and Mexico, and another in Mexico. Florida has at least one representative of each of these five genera. One of them, *Troglocambarus*, is endemic; *Procambarus* is represented by 33 species and subspecies, *Troglocambarus* by one, *Cambarellus* by one, *Orconectes* by one, and *Cambarus* by six. The chart on page 22 presents the progression of taxonomic distinctions within this subfamily made by various students since 1870.

Definition.—The first abdominal segment of the male bears a pair of variously shaped appendages with from two to five terminal elements, one of which always serves as a sperm conduit [the corresponding appendages of the female are reduced or absent]. The podobranchiae of the second and third maxillipeds and of the first three pairs of pereiopods possess "a broad bilobed plaited lamina; the epipodite of the first maxilliped is destitute of branchial filaments; the coxopoditic setae are acute, not hooked, at the end; the telson is commonly divided more or less completely by a transverse suture" (Faxon 1885a: 2). Gills are absent on the last thoracic somite, and there is no bilobed lamina on the podobranchiae of the fourth pair of pereiopods.

THE EVALUATION OF TAXONOMIC CHARACTERS IN THE CAMBARINAE

Considerable emphasis has in the past been given the presence or absence of hooks on the ischiopodites of the second, third, and fourth pereiopods of the male in arriving at taxonomic groupings. These hooks are used by the male in copulation, and are much more strongly developed in the breeding individuals. While this character is extremely useful in many instances, it cannot always be taken as indicative of kinship. *Procambarus pubischelae*, for example, shows marked variation in this regard. In certain portions of its range the individuals have hooks on the ischiopodites of the third and fourth pereiopods, while in other regions hooks are present on only the third pair. While so extreme a variation is known only in the case of this one species, there are several instances in which it is approximated. The following may be taken as an example. *Procambarus geodytes* which is beyond any doubt closely allied to *P. advena* (hooks present on the ischiopodites of only the third pereiopods) has hooks on the ischiopodites of the third and fourth pereiopods, and if some of the former groupings (Faxon 1914, Groups II and III) were followed, this species would be entirely separated from *advena*, and placed in a group where it very obviously does not belong. Both Ortmann (1905a: 95) and Creaser (1931b: 10) have pointed out the independent development (or primitive retention) of hooks on the ischiopodites of the fourth pereiopods in certain species, the relatives of which have hooks on the third pereiopods only.

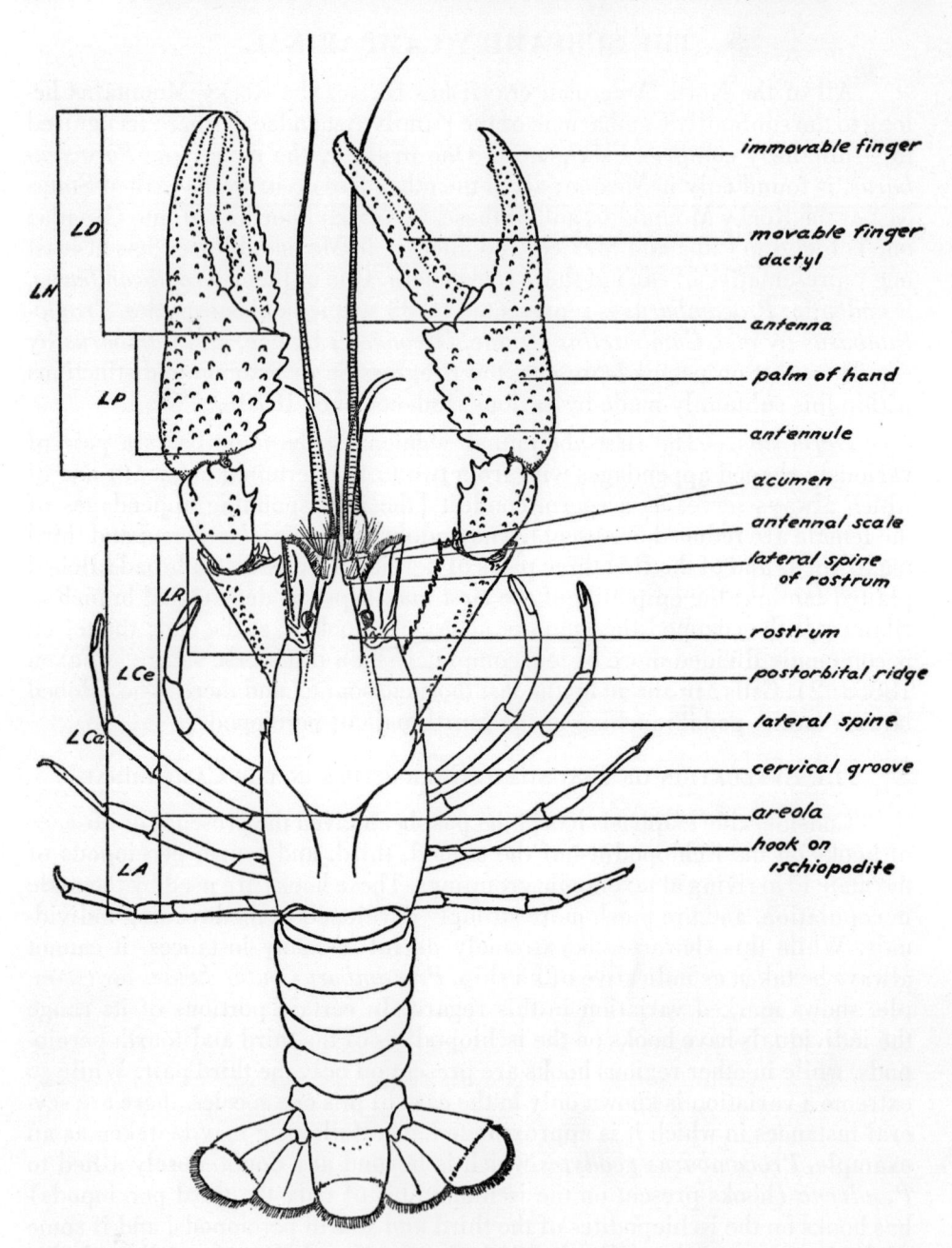

Text Figure b.—Terminology and Methods of Taking Measurements. LH—length of outer margin of hand; LD—length of dactyl; LP—length of palm; LCa—length of carapace; LCe—length of cephalic section of carapace; LR—length of rostrum; LA—length of areola.

Other characters that have been appealed to for establishing species groups appear to be clearly adaptive rather than phylogenetic. The length-breadth ratio of the areola shows a definite correlation with the type of habitat, and in some instances is probably as indicative of habitat as of relationship. For instance most of the crayfishes that occur in flowing water (presumably with a high oxygen content) have a broader and shorter areola, and hence a smaller gill chamber, than do those living in stagnant waters or in lenitic situations. This correlation of gill chamber and habitat reaches its greatest development in the primary burrowers where the areola tends to be extremely narrow or obliterated. Again, the spines on the rostrum, postorbital ridges, and along the cervical groove tend to be reduced or obsolete in the burrowing colonies of species which have more strongly developed spines in populations living in open water. *P. alleni* and *P. paeninsulanus* are excellent examples. Other conspicuous structural characters such as the shape of the chelae and body proportions show evidence of this same adaptive modification.

The first pleopod of the male and the annulus ventralis of the female have proved to be the most reliable of all taxonomic characters, for they appear to show little variation within the species and to be little affected by adaptation to environment.

The adult male crayfish exhibits two distinctly different and usually alternating morphological forms. Previous authors have referred to these as the "first" and "second" form males or as "form I" and "form II", respectively. These two forms are definitely associated with the reproductive cycle. The "first form" is in the breeding stage and may be distinguished by the presence of the corneous condition of the well defined terminal processes of the first pleopod, while in the "second form" male the terminal processes of the first pleopod are not so well defined, are usually blunt, and are never corneous. Moreover, the hooks on the ischiopodites of the pereiopods are strongly developed and sometimes corneous in the first form, whereas in the second form they are usually reduced and seldom corneous.

The new subdivisions and groupings made necessary by the rich accessions of Florida species have consequently been based upon the structure of the first pleopod and the annulus ventralis supplemented by such special characters as have proved to be apropos and useful in grouping or distinguishing between the various species.

The Chief Characters Used in Keys and Descriptions

The first pleopod of the male (Text Fig. c) is of paramount taxonomic importance, and it is necessary to clearly differentiate and distinguish its parts. The anatomical position which is assumed in the following discussion is with the distal end of the appendage extending ventrad. When the appendage is in resting position, i. e., held against the thorax, the side in contact with the thorax is the cephalic surface, and the side viewed in ventral aspect is the cau-

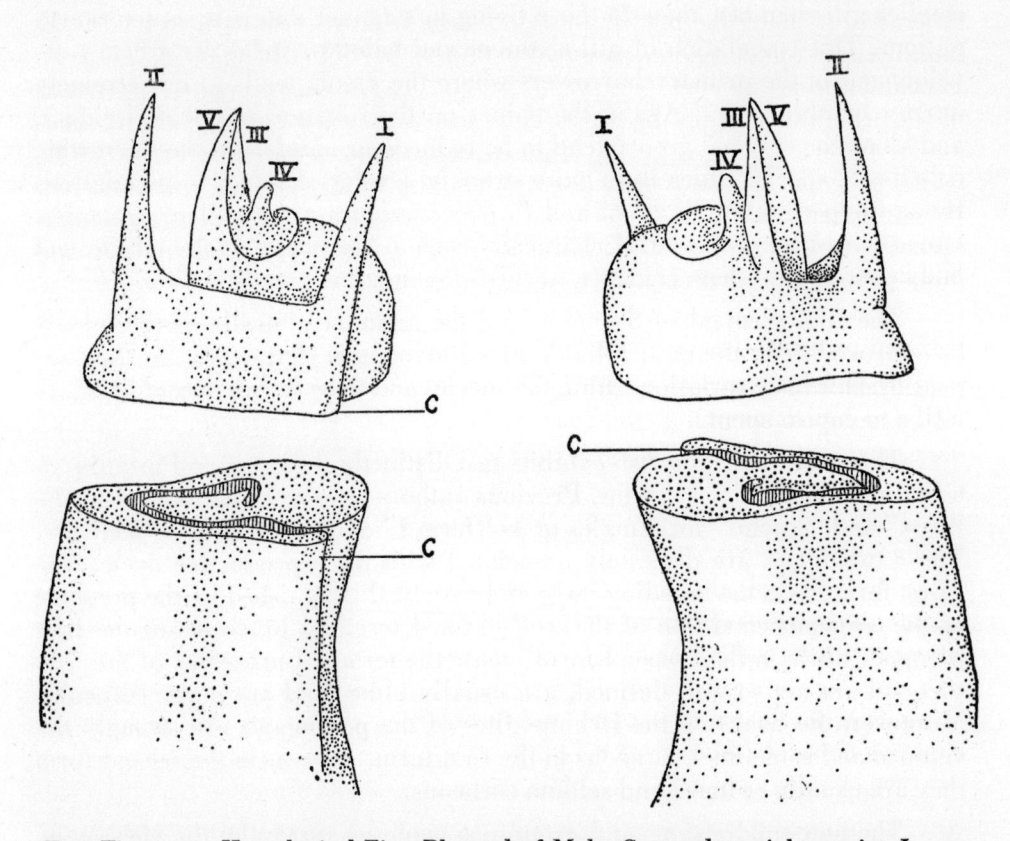

Text Figure c.—Hypothetical First Pleopod of Male. C—caudomesial margin; I—mesial process; II—cephalic process; III—centrocaudal process; IV—caudal process; V—Centrocephalic process; III and V together make up the central projection.

dal surface. The other anatomical directions are obvious when the caudal and cephalic directions are established.

Although there are considerable differences in forms and proportions of the terminal processes of the various species they can all apparently be homologized with a primitive four-parted arrangement (Hobbs, in press-a). Distally the appendage is made up of a rolled plate-like structure which has spines, tubercles, or plate-like projections extending from its distal edge and has its lateral edges folded tightly against the middle portion. The margin "C" is on the caudomesial face of the appendage, and the mesial process "I" extends distad from this margin. As one follows around the perimeter of the appendage he finds a spine projecting from its cephalodistal margin; this is the *cephalic process* "II"; another process, the *centrocaudal* "III", is given off from the distolateral margin, and it makes up the caudal part of the *central projection*. The *caudal process* "IV" projects from the caudolateral, caudal, or caudomesial margin of the appendage, and the cephalic part of the *central projection*, the *centrocephalic* process "V", extends from the inner edge of the roll and projects distally from the center of the appendage.

The terminal processes show considerable variation. Sometimes they are well developed, sometimes one may be reduced, and occasionally one or more may be absent. There is some shifting of the positions of these processes; for example, in most of the members of the *barbatus* (Plates II and III) group the cephalic process arises from the cephalomesial margin of the appendage (Hobbs, in press-a).

The rostrum also shows considerable variation. It may bear lateral spines, the margins may be interrupted without bearing spines, or they may be smooth with no indication of a break between the acumen and the rest of the rostrum. The subrostral ridges are the swollen ridges along the ventrolateral margins of the rostrum, and in some species they extend forward to form the cephalic tip while the rostral ridges fade out before reaching the tip.

The length of the carapace (Text Fig. b) is taken from the tip of the rostrum to the caudal margin of the areola. The length of the areola is taken along the middorsal line, and its width taken at the narrowest part.

There are five pairs of pereiopods (walking legs) of which the chelipeds are the first pair. In the male, hooks are present on the ischiopodites (third segment from the proximal end) of the second and third, third, or third and fourth pereiopods.

The secondary sexual characters of the male which are not mentioned when contrasting the two sexes in description consist of the presence of the above-mentioned hooks on the ischiopodites of the second, third, and fourth pereiopods, and the outgrowths from the bases of the coxopodites (proximal segments) of the fourth and fifth pereiopods.

KEY TO THE FLORIDA CRAYFISHES
(Based on the first form male)

1   Ischiopodite of third maxilliped without teeth on mesial margin. (Only one species of this genus known)............*Troglocambarus maclanei* Hobbs............p. 146

1'   Ischiopodite of third maxilliped with teeth on mesial margin......2

2 (1')   First pleopod of male with two terminal apices that are either straight, divergent, or gently curved; never bent so much as at a right angle to the main shaft. (Only one species, which belongs to the monotypic subgenus *Faxonella*, known in Florida)............*Orconectes clypeata* (Hay)............p. 154

2'   First pleopod of male terminating in two or more parts; if two, these are bent caudad at right angles to the main shaft............3

3 (2')   Hooks on ischiopodites of second and third pereiopods of male. (Only one species of this genus known in Florida)............*Cambarellus schmitti* sp. nov.............p. 149

3'   Hooks on ischiopodites of third, or third and fourth pereiopods...4

4 (3')   First pleopod of male terminating in two blade-like or bulbous processes that are always bent at right angles to the main shaft ....*Genus Cambarus*............40

4'   First pleopod of male terminating in three or more processes that are corneous, spiniform, plate-like, or blunt............*Genus Procambarus*............5

5 (4')   Hooks usually present on ischiopodites of third pereiopods only; if present on both third and fourth, then margins of rostrum uninterrupted (interrupted in some specimens of *alleni*, an occasional specimen of *latipleurum,* and all specimens of the cavernicolous *acherontis*; in *alleni* and *acherontis* the areola is long and narrow, and the hooks on the ischiopodites of the fourth pereiopods are bituberculate; in *latipleurum* the spines on the rostrum are minute and almost at the apex of the rostrum)......6

5'   Hooks on ischiopodites of both third and fourth pereiopods; margins of rostrum interrupted and often bearing lateral spines ....(Blandingii Section)............24

6 (5)   Eyes reduced; carapace without pigment (Acherontis Section with a single species)............*P. acherontis* (Lönnberg)............p. 91

6'   Eyes well developed; carapace pigmented............7

7 (6')   Chela flattened; inner margin of palm bearing a single, well defined, cristiform row of tubercles............(Advena Section)......19

7'   Chela subovate, not strongly compressed; inner margin of palm never bearing a single cristiform row of tubercles, either barbate or bearing several irregular rows of tubercles which are usually small............(Barbatus Section)............8

8 (7')   Hooks on ischiopodites of third and fourth pereiopods; hooks on fourth bituberculate (Alleni Group with a single species)............*P. alleni* (Faxon)............p. 69

8'	Hooks on ischiopodites of third, or of third and fourth pereiopods; hooks on fourth simple............(Barbatus Group)............9	
9 (8')	Hooks on ischiopodites of third and fourth pereiopods............10	
9'	Hooks on ischiopodites of third pereiopods only............17	
10 (9)	Palm of chela barbate along inner margin............16	
10'	Palm of chela not barbate along inner margin............11	
11 (10')	Cephalic process of first pleopod arising from cephalic surface............*P. shermani* sp. nov.............p. 61	
11'	Cephalic process of first pleopod arising from mesial surface............12	
12 (11')	Mesial process of first pleopod conspicuously large, subspatulate, bent caudad at an angle of 45-90 degrees to the main shaft of the appendage............*P. kilbyi* (Hobbs)............p. 64	
12'	Mesial process of first pleopod never subspatulate; directed distad or caudodistad but not so much as at a 45 degree angle to the main shaft............13	
13 (12')	Caudal process of first pleopod thumb-like, or if not thumb-like the cephalic process curved caudad............14	
13'	Caudal process of first pleopod not thumb-like, forming a laterally compressed blade-like structure along the caudolateral tip of the appendage; cephalic process straight or curved cephalad............15	
14 (13)	Cephalic process of first pleopod straight; mesial process acute and directed caudodistad............*P. apalachicolae* sp. nov. p. 55	
14'	Cephalic process of first pleopod curved caudad; mesial process subspiculiform and directed distad but curved cephalad............*P. latipleurum* sp. nov.............p. 52	
15 (13')	Coxopodite of fourth pereiopod with no prominent outgrowth; outgrowth on that of fifth pereiopod simple. Opposable margin of dactyl of chela with only the faintest indication of an excision. Inner margin of palm of chela never barbate............*P. econfinae* sp. nov.............p. 49	
15'	Coxopodite of fourth pereiopod with a single large tubercle; outgrowth on that of fifth with two or three emarginations. Opposable margin of dactyl of chela with a distinct excision. Inner margin of palm of chela usually barbate............*P. escambiensis* sp. nov.............p. 46	
16 (10)	Cephalodistal surface of first pleopod with a rounded prominence; cephalic process small and acute and arising from the mesial surface of the appendage; caudal process rounded and noncorneous............*P. pubischelae* sp. nov.............p. 41	
16'	Cephalodistal surface of first pleopod without a rounded prominence, cephalic process blade-like and arising from the cephalic surface of the appendage; caudal process with sharp corneous edges............*P. shermani* sp. nov.............p. 61	

17 (9')	Inner margin of palm of chela not barbate............*P. rathbunae* (Hobbs)............p. 59
17'	Inner margin of palm of chela barbate............18
18 (17')	Caudal process of first pleopod forming a corneous blade-like structure along the caudolateral tip of the appendage............ *P. hubbelli* (Hobbs)............p. 67
18'	Caudal process of first pleopod rounded and non-corneous............ *P. pubischelae* sp. nov............p. 41
19 (7)	Hooks on ischiopodites of both third and fourth pereiopods............ *P. geodytes* sp. nov............p. 80
19'	Hooks on ischiopodites of third pereiopods only............20
20 (19')	First pleopod of male with the plate-like, corneous fan formed by the central projection directed aross the cephalic surface of the tip............22
20'	First pleopod of male with the plate-like, corneous fan formed by the central projection directed cephalocaudad............21
21 (20')	Cephalic process of first pleopod present as a vestige............*P. advena* (LeConte)............p. 75
21'	Cephalic process of first pleopod absent............*P. pygmaeus* sp. nov............p. 83
22 (20)	The plate-like central projection of the first pleopod directed obliquely caudad across the cephalic border of the tip; cephalic process present............*P. rogersi ochlocknensis* Hobbs............p. 89
22'	The plate-like central projection of the first pleopod directed laterad across the cephalic border of the tip; cephalic process present or absent............23
23 (22')	Caudal process of first pleopod bent, but not so much as at a right angle to the main shaft of the appendage............*P. rogersi campestris* Hobbs............p. 90
23'	Caudal process of first pleopod bent at a right angle to the main shaft of the appendage............*P. rogersi rogersi* (Hobbs) p. 89
24 (5')	Eyes reduced; carapace without pigment............(Pictus Group, in part)............29
24'	Eyes well developed; carapace pigmented............25
25 (24')	First pleopod bearing a hump on cephalic margin (Clarkii Subgroup of Blandingii Group)............35
25'	First pleopod never bearing a hump on cephalic margin............26
26 (25')	Chela bearing a row of seven or less tubercles along mesial margin of palm (if more, then a median carina is present on upper surface of rostrum)............Spiculifer Group............39
26'	Chela bearing a row of eight or more or several rows of tubercles along mesial margin of palm, or inner margin of palm smooth. Never with a median carina present on rostrum............27
27 (26')	Areola broad and short. Cephalic process of first pleopod spiniform and clearly extending beyond the main body of the appendage............(Pictus Group, in part)............29

27'		Areola narrow and long. Cephalic process of first pleopod sometimes spiniform, but if so extends barely distad of the main body of the appendage (Blandingii Group)_____28
28	(27')	First pleopod with all four terminal apices strongly developed, the cephalic and caudal process and the central projection all corneous and blade-like_____Blandingii Subgroup_____34
28'		First pleopod never with all four terminal apices strongly developed, and the cephalic and caudal processes and the central projection never all blade-like_____(Fallax-Evermanni Subgroups)_____36
29	(24 and 27)	Carapace pigmented; eyes well developed_____32
29'		Carapace without pigment; eyes reduced_____30
30	(29')	Eye with small pigment spot_____*P. lucifugus alachua* (Hobbs)_____p. 136
30'		Eye without small pigment spot_____31
31	(30')	Rostrum broadest distad of base_____*P. lucifugus lucifugus* (Hobbs)_____p. 134
31'		Rostrum broadest at base_____*P. pallidus* (Hobbs)_____p. 139
32	(29)	Length of inner margin of palm of chela greater than length of dactyl; acumen of rostrum as long as or longer than rest of rostrum_____*P. youngi* sp. nov._____p. 131
32'		Length of inner margin of palm of chela less than length of dactyl; acumen of rostrum much shorter than rest of rostrum_____33
33	(32')	Central projection of first pleopod extends distad to tip of cephalic process; both spiculiform; cephalic process directed distad _____*P. seminolae* sp. nov._____p. 142
33'		Central projection of first pleopod extends distad but never so far as tip of cephalic process; neither spiculiform; cephalic process directed at about a 45 degree angle to the main shaft_____ *P. pictus* (Hobbs)_____p. 130
34	(28)	Length of rostrum greater than length of areola; a dark stripe on either side of thoracic portion of carapace that persists even in alcohol_____*P. bivittatus* sp. nov._____p. 96
34'		Length of rostrum less than length of areola; no dark stripe on thoracic portion of carapace_____*P. blandingii acutus* (Girard)_____p. 94
35	(25)	Mesial process of first pleopod directed distad. Caudal process forms an undulating blade across the caudolateral surface (in ventral view). Cephalocaudal dimension of cephalic process greater than length_____*P. okaloosae* sp. nov._____p. 100
35'		Mesial process of first pleopod directed caudad. In ventral view caudal process forms a narrow blade closely applied to the central projection. Cephalocaudal dimension of cephalic process less than length_____*P. paeninsulanus* (Faxon)_____p. 104
36	(28')	Mesial process of first pleopod blade-like_____*P. fallax* (Hagen)_____p. 111

36'	Mesial process of first pleopod spiniform 37
37 (36')	Caudal process of first pleopod long and blade-like, extending distad almost to tip of central projection .... *P. evermanni* (Faxon) .... p. 107
37'	Caudal process of first pleopod vestigial, hardly discernible; central projection reduced to a minute corneous spine or blade extending from the center of the appendage 38
38 (37')	Distal end of first pleopod bent at about a 45 degree angle to the main shaft; shoulder on cephalodistal surface rounded .... *P. leonensis* Hobbs .... p. 114
38'	Distal end of first pleopod only slightly bent; shoulder on cephalodistal surface angular .... *P. pycnogonopodus* Hobbs .... p. 117
39 (26)	Rostrum with slight median carina .... *P. versutus* (Hagen) .... p. 126
39'	Rostrum without median carina .... *P. spiculifer* (LeConte) .... p. 119
40 (4)	Areola obliterated at midlength .... (Diogenes Section) 43
40'	Areola broad or narrow, but never obliterated at midlength .... (Bartonii Section) 41
41 (40')	Eyes reduced; eyes and carapace without pigment .... *C. cryptodytes* Hobbs .... p. 162
41'	Eyes well developed; both eyes and carapace pigmented 42
42 (41')	Areola with three or four punctations in narrowest part .... *C. latimanus* (LeConte) .... p. 158
42'	Areola with one or two punctations in narrowest part .... *C. floridanus* Hobbs .... p. 161
43 (40)	Rostrum lanceolate or with acumen distinct, well set off, deeply excavate. No cristiform row of tubercles along inner margin of palm of chela .... *C. diogenes diogenes* (Girard) .... p. 164
43'	Rostrum subovate, acumen never sharply set off, deeply or shallowly excavate. A more or less distinct cristiform row of tubercles along inner margin of palm of chela 44
44 (43')	Rostrum short; directed ventrad; margins converging from base. The cristiform row of tubercles along inner margin of palm acute and well marked. Cephalomesial margin of antennal scale slanting or gently rounded .... *C. byersi* Hobbs p. 167
44'	Rostrum long; directed ventrad, but not so greatly as in *byersi*; margins subparallel for some distance at base. The cristiform row of tubercles along inner margin of palm of chela less acute than in *byersi*. Cephalomesial margin of antennal scale angulate .... *C. species incertis* .... p. 168

Genus PROCAMBARUS Ortmann

Groups I and III (in part) of *Cambarus*, Hagen 1870, Ill. Cat. Mus. Comp. Zool., Harvard Coll., No. 3: 32-55, 74, 84, 87-88.

Groups I and II of *Cambarus*, Faxon 1885a, Mem. Mus. Comp. Zool., Harvard Coll., 10 (4): 17, 47.

Subgenus *Cambarus*, Ortmann 1905a, Proc. Amer. Philos. Soc., 44 (180): 96-106.

Subgenus *Procambarus*, Ortmann 1905c, Ann. Carnegie Mus., 3 (3): 437.

Subgenus *Ortmannicus*, Fowler 1911, Ann. Rep. New Jersey Mus. 1911, Part II: 340.

Groups I, II, III of *Cambarus*, Faxon 1914, Mem. Mus. Comp. Zool., Harvard Coll., 40 (8): 410-414.

Subgenus *Cambarellus*, Creaser 1933b, Occ. Pap. Mus. Zool., Univ. Mich., (275): 21.

Genus *Procambarus*, Hobbs (in press-c), Amer. Mid. Nat.

A historical summary of the taxonomic treatment of the group by former authors has been published elsewhere (Hobbs, in press-c). A table from this paper (p. 22) shows the synonymy involved.

Diagnosis.—"First pleopod of first form male terminating in from two to five distinct parts which may be truncate, plate-like or spiniform. Shoulder present or absent on cephalic surface of distal third. If the pleopod terminates in only two parts this shoulder is always present. Hooks present on the ischiopodites of the third or of the third and fourth pereiopods in the male. Third maxillipeds of normal size bearing a row of teeth along the inner margin of the ischiopodite" Hobbs (in press-c).

The geographic range of this genus is undoubtedly more extensive than that of any of the others in the Subfamily Cambarinae, extending from Wisconsin and Colorado to Guatemala and eastward to the Atlantic seaboard from New York to Florida.

Except in the mountainous regions one or more representatives of the genus can be found in any body of freshwater, or nearly any low, poorly drained area, but it reaches its greatest abundance in the coastal plains of the southern and southeastern United States.

In Florida *Procambarus* is represented by 33 species and subspecies and extends throughout the state from the western extremity of the northwestern panhandle to Big Pine Key, off the southern tip of the mainland.

THE BARBATUS SECTION

Diagnosis.—Cephalodistal surface of the first pleopod of first form male terminates in a ridge (sharp or truncate) or a knob-like prominence which is distinctly not a part of one of the terminal processes; mesial process always directed distad unless spatulate; cephalic process, when present, always extends distad from mesial surface (except in *shermani*); central pro-

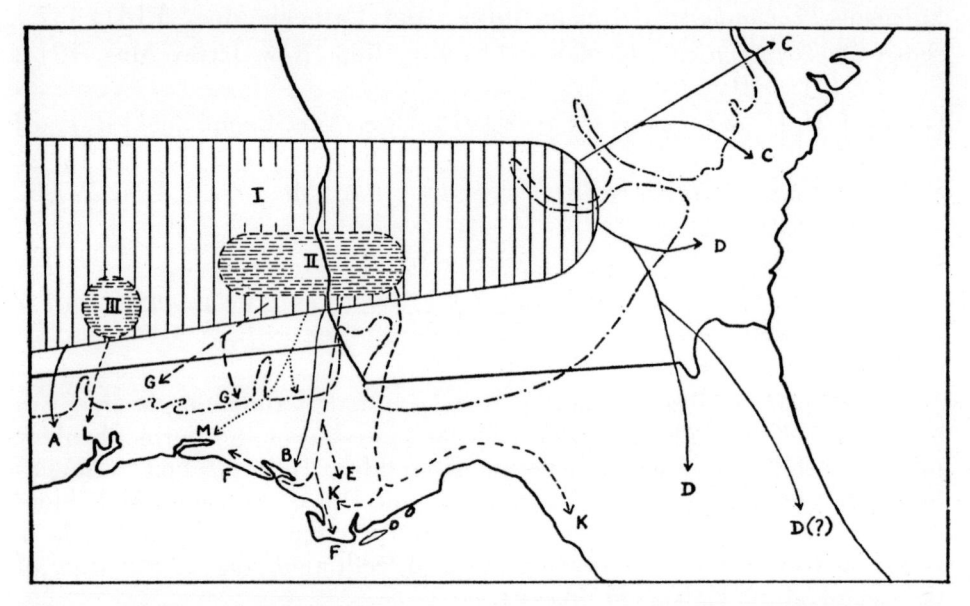

Map 3.—Migrations of the Barbatus Group. The dot-dash line shows the Sunderland Shore Line of early Pleistocene; the shaded portions indicate the areas inhabited by the ancestral stock during early Pleistocene. The earliest wave of migration probably came from the general region designated by I; later migrations appear to have been derived from the more circumscribed centers, shown as II and III. The letters indicate the general regions in which the modern species occur, and the lines represent the probable paths taken by their ancestors. The species A—*escambiensis*, B—*econfinae*, C—*barbatus*, D—*pubischelae*, E—*latipleurum*, F—*apalachicolae*, and G—*rathbunae* are members of the *barbatus* subgroup, while K—*kilbyi*, L—*shermani*, and M—*hubbelli* belong to monotypic subgroups, designated respectively by the names of the species.

jection never decidedly the most conspicuous terminal element. Hooks present on ischiopodites of third, or third and fourth pereiopods.

Fourteen of the known species of *Procambarus* are referable to the *barbatus* section: *barbatus, pubischelae, escambiensis, econfinae, latipleurum, apalachicolae, rathbunae, shermani, kilbyi, hubbelli, simulans, gracilis, hagenianus,* and *alleni*. Ten of these occur in Florida, and one or more has been recorded from New Mexico, Colorado, Texas, Oklahoma, Kansas, Arkansas, Missouri, Iowa, Wisconsin, Illinois, Mississippi, Alabama, and Georgia.

The Barbatus Section may be divided into the following groups and subgroups:

BARBATUS SECTION
    Barbatus Group
        Barbatus Subgroup
        Shermani Subgroup
        Kilbyi Subgroup
        Hubbelli Subgroup
    Alleni Group

The Barbatus Group

Diagnosis.—First pleopod of first form male terminates in four distinct parts and bears no decided hump or shoulder on cephalic margin. Areola relatively short and broad; no lateral spines present on rostrum nor are margins interrupted (except occasionally in *latipleurum*); hooks present on ischiopodites of third, or third and fourth pereiopods.

The *barbatus* group comprises an assemblage of secondary burrowing species which are almost confined to the flatwoods of the coastal terraces (or adjoining flatwoods) from the Perdido River drainage system eastward into South Carolina, and southward to Flagler, Alachua, and Levy counties in Florida. It should be noted that within these boundaries the actual range of the group is broken into distinctly discontinuous areas. (Map 4).

Of the ten species which belong to the *barbatus* group *(barbatus, pubischelae, escambiensis, econfinae, latipleurum, apalachicolae, rathbunae, shermani, kilbyi,* and *hubbelli)* all except the first occur in Florida, and the last seven are thus far known only from this state. It is probable, however, that further collecting will extend the range of *escambiensis, rathbunae, shermani,* and *hubbelli* into Alabama.

The *barbatus* subgroup appears to me to be a natural assemblage of related species, and the relationships indicated by morphological criteria are directly correlated with the existing pattern of geographic distribution. It is thus possible to postulate the paths of migration by which the progenitors of the present *barbatus* subgroup occupied the areas where they are now established.

Although *shermani, kilbyi,* and *hubbelli* are closely related to the members of the *barbatus* subgroup, their anatomical peculiarities are so distinct from the other members of the *barbatus* group and from each other that it

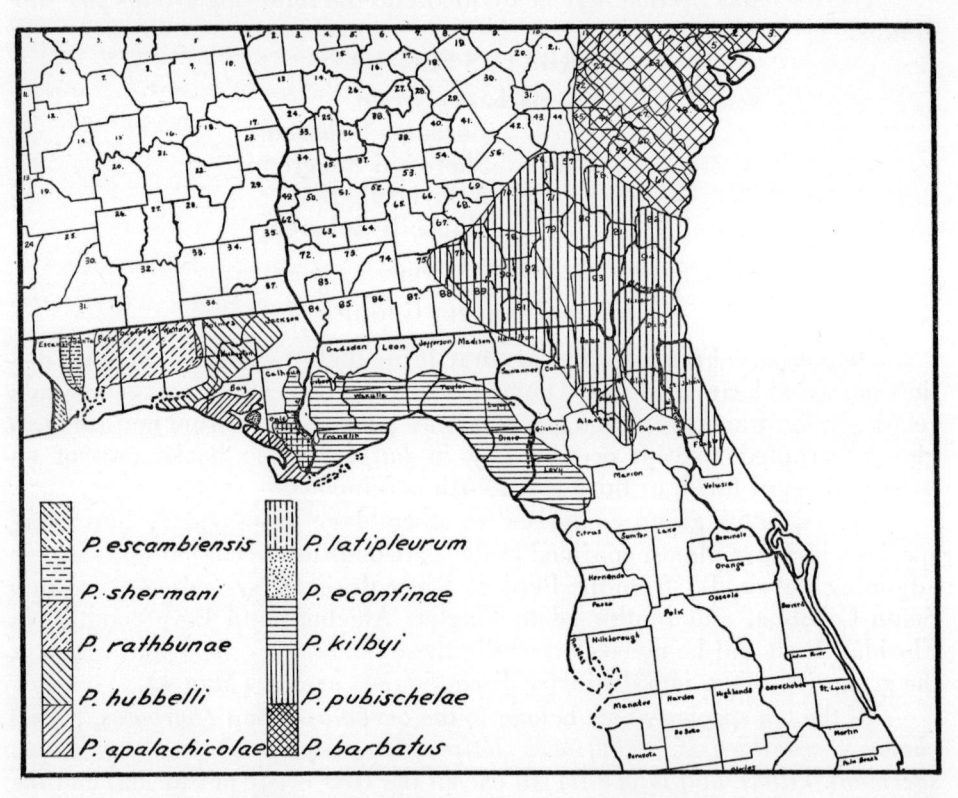

Map 4.—Distribution of the Barbatus Group.

seems best to erect separate subgroups for their reception. As will be pointed out below they probably originated from different stocks of the *barbatus* group, independently.

The present distribution of the *barbatus* group is most acceptably explained on the assumption that its forerunners occupied the gulf coastal region (I) as indicated (Map 3), comparatively early in the Pleistocene emergence of the coastal plain; a southward and eastward migration took place that resulted in the isolation of at least three groups: a western (A), a central (B), and a northeastern (C and D) which occupied respectively the regions of the drainages of the Perdido, Flint-Chattahoochee (or Choctawhatchee), and Altamaha rivers. Perhaps a fourth group (L) which took a different southward path along the Escambia should be considered here; however, I am inclined to believe that this migration should be referred to a later period.

The western group progressed southward along the Perdido River where it was isolated from the central group by the high, rolling country in Okaloosa and Walton counties. This stock gave rise to *P. escambiensis* (A).

The central stock progressed southward either along the Flint-Chattahoochee or along the Choctawhatchee drainages giving rise to *econfinae* (B).

The third stock, moving eastward, diverged, part (C) to cross the Altamaha and Savannah rivers and part (B) to extend southward into Florida. The part that moved north of the Altamaha gave rise to *P. barbatus* while that which moved southward was ancestral to *P. pubischelae* in Georgia south of the Altamaha River and in northeastern Florida.

Following the first wave of migration which resulted in the isolation of *escambiensis, econfinae, barbatus,* and *pubischelae,* a second, and possibly a third, wave of migration is strongly indicated. In the second region the migration came from two separate but more restricted regions (labelled II and III on the map), one in the vicinity of the headwaters of the Escambia River in southern Alabama and the other in southeastern Alabama and southwestern Georgia. The stock from the more western area migrated southward along the Escambia River to give rise to *P. shermani* (E).[7]

The group which occupied Area II in southeastern Alabama and southwestern Georgia extended southward along three separate routes. The more western one (G) extended in a southwesterly direction and reaching the Blackwater and Yellow river systems gave rise to *P. rathbunae.* Another line of migration followed the west side of the embayment of the Flint River into the coastal region, and there segregated to form *P. apalachicolae* (F) and *P. latipleurum* (E). Still farther east another stock pushed southward to give rise to *P. kilbyi* (K) which followed the coast southward as far as Levy County, and westward across the Apalachicola River into Calhoun, Franklin, and Gulf counties.

---

[7]This species is particularly interesting in that it possesses characters which are remarkably like those of the *blandingii* section. Further consideration of this species and its relationships is given below.

The origin and time of migration of the stock from which *P. hubbelli* was derived are more questionable. On morphological grounds *hubbelli* is evidently close to the original Pro-barbatus stock and perhaps should be referred to the earliest migration across the Sunderland Terrace, but the relationship of its present range to that of other species in the central portion of the panhandle seems to indicate a comparatively late, delayed migration from the region of Area II.

All of the species derived during the first period of migration share a number of peculiar characters, and with the exception of *econfinae*, which almost certainly has its closest affinities with *escambiensis*, all have first form males that show a heavy beard along the inner margin of the palm of the chela. Except for *shermani*, those species derived from subsequent migrations have a naked chela, and the pleopods of all of this group (except for *shermani* and *kilbyi*) are remarkably similar. *P. shermani* is unique in so many ways in this assemblage that it may well have been derived from a stock already differentiated from that which provided the main body of later migrants. *P. hubbelli* is likewise a disjunct species in this group, although it has barbate chelae.

It is interesting to note how definitely the present ranges of the members of the *barbatus* group illustrate Matthews' "Center Fire" hypothesis of distribution. There is little doubt that this group had its distribution from the older region of the southeastern coastal plain, and now we find the periphery of the present range occupied by the earliest segregates from the original stock, while the region between the center of dispersal and the periphery occupied by later migrants from the center of dispersal. The older species are *barbatus* and *pubischelae* along the eastern seaboard, *econfinae* in a small area around Panama City (completely surrounded by members of the later migration), and *escambiensis* on the western periphery of the range, along the Perdido River; the more recent forms, particularly *rathbunae*, *apalachicolae*, and *latipleurum* have filled in the gaps, and are established farther inland. The latter two particularly indicate a more recent advent into the Panama City region as evidenced by their complete encirclement of *P. econfinae*.

The first pleopods of the *barbatus* group, especially those of *P. barbatus*, very closely approximate the form of the appendages in *P. mexicanus*, and it is possible that *barbatus* more nearly represents the ancestral stock than do any of the other members of the group.

The Barbatus Subgroup

Diagnosis.—Cephalodistal surface of first pleopod of first form male generally compressed and bears a row of setae; mesial process typically spiniform, occasionally flattened, but more often cylindrical, extends distad beyond the rest of the appendage; cephalic process, small and corneous, has shifted mesially and arises from the distomesial margin (except in *shermani* where it arises from the cephalic surface); central projection markedly variable, corneous, and very small, generally forms a triangular plate that is flattened cephalocaudad; caudal process usually large and terminates in either a knob, a distally flattened prominence, a corneous blade-like structure directed cephalocaudad, or a subtriangular, non-corneous process. In most instances the distal portion of the appendage is bent more or less caudad. The areola is broad or relatively broad; the margin of the rostrum is toothless and uninterrupted (except in occasional specimens of *latipleurum*); the ischiopodites of the third pereiopods bear hooks while those of the fourth may be with or without hooks; the inner margin of the palm of the chelae of the male may or may not be barbate.

The members of the *barbatus* subgroup commonly inhabit flatwoods, and generally live in temporary bodies of water. When the pools dry up the crayfishes dig into the wet soil of the old pool bottoms and construct rather simple burrows which may range from six inches to three feet in depth, depending on the soil type and the length and severity of the period of drought. Specimens are occasionally found in the smaller streams of the flatwoods, but these stream dwellers also construct burrows at the breeding season. I have never collected specimens from a large body of water, and there is some evidence that large rivers act as a barrier to certain of the species.

The range of the *barbatus* subgroup is confined to the southeastern coastal plain where it is separated into an eastern and western portion. The eastern portion extends northward into the southeastern part of South Carolina; on the northwest the boundary is marked by the edge of the Tifton Upland; on the west (south of the Tifton Upland) by the comparatively high, well drained region in Hamilton, Columbia, Union, and Alachua counties, Florida. The southern part of the range extends into northern Florida, southward to Alachua and Flagler counties. The western portion of the range has its eastern boundary along the Apalachicola River; its northern boundary is the northern limit of the coastal flatwoods in Gulf, Bay, Walton, Okaloosa, Santa Rosa, and Escambia counties, Florida; its western boundary is in the vicinity of the Perdido River.

Generally in Florida these flatwoods extend hardly more than 25 miles from the coast, although Flomaton, Escambia County, Alabama, where *P. escambiensis* has been reported, lies approximately 30 miles from the coast.

Procambarus barbatus (Faxon)

Plate II, Figs. 1-5; Maps 3, 4

Astacus penicillatus LeConte (not *Astacus penicillatus* Olivier 1791) 1856, Proc. Acad. Nat. Sci. Philad., 7: 401.

Cambarus penicillatus Hagen 1870: 16, 31-33, 53-54, 55, 97, 100, 106, 107, 108, Pl. I, figs. 93, 94; Pl. III, fig. 149; Faxon 1884 (part): 138; Faxon 1885a (part): 18, 36-38, 39, 158, 173.

Cambarus barbatus Faxon 1890: 621, new name to replace *penicillatus* LeConte 1856, preoccupied by *penicillatus* Olivier 1791 (among Faxon's neotypes are specimens from the Escambia River at Flomaton, Alabama, herein described as *P. escambiensis* Hobbs); Harris 1903 (part): 58, 72, 137, 138, 150, 152, 153, 154; Hobbs 1940a: 389, 410, 414, 418; Hobbs 1940b: 3; Hobbs (in press-a).

Procambarus barbatus Hobbs (in press-c).

Diagnosis.—Rostrum without lateral spines; areola relatively broad with three to five punctations in narrowest part; male with hooks on ischiopodites of third and fourth pereiopods; inner margin of palm of chela of first form male usually bearded within (bearing a heavy growth of plumose setae); postorbital ridges terminating cephalad without spines; no lateral spines present on carapace. First pleopod of male, form I, reaching coxopodite of third pereiopod and terminating in four parts; mesial process, the largest of the four, corneous, subspiculiform, and extending distad and slightly recurved cephalad; cephalic process small, corneous, and arising from mesial surface of appendage, is curved and projects from beneath distal portion of central projection; caudal process, hardly discernible from the truncate portion, somewhat flattened, and forming the caudodistal portion of the outer part; central projection small, corneous, triangular, and arising from central portion of tip, is flattened cephalocaudad and directed cephalodistad, somewhat overhanging the basal portion of the cephalic process. Cephalic margin of main shaft near tip bearing a cluster of setae which forms a single row as it approaches the terminal processes. Annulus ventralis subrectangular with prominent tubercles on each cephalolateral margin; sinus originating near center of annulus, slightly dextrad of midventral line, curves gently dextrad, then sharply sinistrad to cross midventral line where it makes a hairpin turn almost back to midventral line; thence caudad to caudal margin.

Remarks.—Faxon (1885a: 38) states, "LeConte does not specify from what part of Georgia his specimens come, nor is the locality of the Georgia specimens in the Museum of Comparative Zoology any more precisely indicated." LeConte's types of this species are unknown. According to Faxon (1890: 621) "... LeConte's specific name *penicillatus* must be rejected, since it has been used previously by Olivier (Encyc. Meth., Hist. Nat. des Insectes, VI, 1791, p. 343), in combination with the same generic name, for another animal (*Palinurus penicillatus* of recent authors)." In this paper Faxon proposed the name, *Cambarus barbatus*, to replace *Cambarus penicillatus*. Faxon (1914: 414) cites a first form male from Georgia, but without further data, (M. C. Z. no. 279) as his type specimen, and several specimens from the Escambia River, Flomaton, Alabama (M. C. Z. no. 3845) as paratypes. My assignment of the name *barbatus* is based on study of Faxon's Neotype (M. C. Z. no. 279); his "paratypes" belong to another species, *Procambarus escambiensis*, herein described.

I have made camera lucida drawings of Faxon's neotype and "paratypes" at M. C. Z., and since that time have collected in several counties in southeastern Georgia and South Carolina. Among my specimens from Liberty County, Georgia, there is a first form male which resembles the neotype very closely. In addition, I also have a large number of specimens collected north of the Altamaha River in Georgia and South Carolina which belong to this species. Collections made in Georgia south of the Altamaha River have disclosed a very closely related form but not *P. barbatus*. I believe it likely that Faxon's neotype was collected from the region between the Altamaha and Savannah rivers in the coastal plain.

Although this species does not occur in Florida it is included here for two reasons. It has been confused with a very closely allied species which occurs in Florida, and it is also the only known non-Floridian member of the *barbatus* subgroup, unless *Procambarus pearsei* (Creaser) is also a member of this group.

Distribution.—GEORGIA—Bulloch, Effingham, Jenkins, Liberty, Long, and Screven counties. SOUTH CAROLINA—Hampton County (Hobbs 1940b: 3). Faxon cites the following records with a query: South Carolina (?); Charleston (?); Mississippi (?); Eastern (?).

All of the records cited above, except where otherwise indicated, are new; heretofore the only data given has been "Lower Georgia" or Georgia.

Procambarus pubischelae, sp. nov.

Plate II, Figs. 6-10; Plate XIII; Maps 3, 4.

Diagnosis.—Rostrum without lateral spines; areola relatively broad with three to five punctations in narrowest part; male with hooks on ischiopodites of third only, or third and fourth pereiopods; palm of chela of first form male usually bearded; postorbital ridges terminating cephalad without spines; no lateral spines present on carapace. First pleopod of male, form I, reaching coxopodite of second pereiopod and terminating in four distinct parts; mesial process corneous and subspiculiform; cephalic process corneous, small, and spiniform; caudal process noncorneous and truncate; central projection corneous, subtriangular, and compressed cephalocaudad. Cephalodistal setae-bearing prominence subangular. Annulus ventralis subrectangular with a triangular lateral projection on each side; excavate cephalad with tuberculate prominences laterad; sinus originating dextrad of midventral line, bends sharply sinistrad, and runs laterad to cross the midventral line where it forms an arc and terminates just before cutting the caudal margin sinistrad of middle.

Holotypic Male, Form I.—Body subovate, compressed laterally. Abdomen narrower than thorax (1.21-1.35 cm. in widest parts respectively).

Width and depth of carapace subequal in region of caudodorsal margin of cervical groove. Greatest width of carapace slightly caudad of caudodorsal margin of cervical groove.

Areola of moderate width (4.68 times longer than wide); with four punctations in narrowest part; sides not parallel. Cephalic section of carapace about 1.99 times as long as areola (length of areola 33.5% of entire length of carapace).

Rostrum flattened above, almost reaching base of distal segment of peduncle of antennule; margins converging to tip; no lateral spines present. Upper surface of rostrum punctate; marginal ridges rather high and sharp. Subrostral ridge well defined but evident in dorsal view for only a short distance from base.

Postorbital ridges well defined, merging rather abruptly cephalad into carapace without forming a tubercle or spine. Suborbital angle rounded, only moderately prominent; branchiostegal spine well developed; no lateral spines on sides of carapace, a few small tubercles present instead. Surface of carapace punctate dorsad, weakly granulate laterad.

Abdomen longer than thorax (3.0-2.6 cm.).

Cephalic section of telson with two spines in each caudolateral corner.

Epistome subovate with a single emargination cephalad.

Antennule of usual form; a spine present on ventromesial side of basal segment.

Antennae reaching caudad to fourth abdominal segment. Antennal scale broad, broadest in middle; inner margin evenly rounded; spine on outer margin moderately strong.

Chela subovate, compressed dorsoventrally, of moderate length, and broad. Hand entirely tuberculate except on ventrolateral surface which bears setiferous punctations. Inner margin of palm with a heavy beard of plumose setae. A distinct ridge present on both fingers. Fingers not gaping. Opposable margin of dactyl with a row of 11 rounded tubercles, and with sparsely scattered denticles along distal half. Lateral margin of dactyl with a row of six subsquamous, ciliated tubercles along basal half. Upper surface with a single submedian ridge at the base of which are a few small scattered tubercles, flanked with scattered setiferous punctations distad. Lower surface also with a distinct median ridge and scattered setiferous punctations. Opposable margin of immovable finger with six rounded tubercles and with minute denticles scattered along entire length. Lateral margin with a weak ridge. Upper surface tuberculate proximad and punctate distad with a submedian ridge. Lower surface with a distinct submedian ridge and scattered punctations. A single row of punctations bearing long, stiff setae present below opposable margins of both fingers.

Carpus of first pereiopod longer than wide (.80-.56 cm.); shorter than inner margin of palm of chela (.88 cm.); a distinct longitudinal groove above; dorsomesial and mesial surfaces tuberculate; otherwise sparsely punctate. Two larger tubercles along midmesial surface, and two along cephalic dorsomesial margin. Three large tubercles present in a row running obliquely from lower cephalomesial surface to cephalic margin of ventral surface.

Merus of first pereiopod punctate mesiad and laterad. A group of scattered tubercles present on upper surface, more numerous distad. Lower surface with two rows of tubercles, an outer one of about nine which are smaller in size than the inner row of about 14, most of which are spike-like.

Hooks present on ischiopodites of third and fourth pereiopods. Bases of coxopodites of fourth and fifth pereiopods with strong outgrowths directed caudoventrad and cephaloventrad respectively.

First pleopod reaching base of second pereiopod when abdomen is flexed; terminal portion of appendage with a tuft of setae on cephalic surface. Appendage ending in four distinct parts. Mesial process subspiculiform and corneous, extends beyond the rest of the terminal processes in a cephalodistal direction. Cephalic process, small and corneous, arises from the mesial surface and extends almost parallel to the mesial process. Caudal process truncate, noncorneous, heavy and slightly excavate distad, makes up the caudolateral portion of the tip. Central projection small, corneous, subtriangular, flattened cephalocaudad, and directed cephaloventrad. Cephalodistal portion of the appendage while rounded in lateral aspect is somewhat dome-shaped in contrast to the gradual sloping appearance seen in *Procambarus barbatus*.

Male, Form II.—Differs from the first form male in the following respects: inner margin of palm of chela never bearded; all spines and tubercles greatly reduced in size and somewhat in number; hooks on ischiopodites of third and fourth pereiopods very small; no distinction between the caudal process and the central projection of the first pleopod, no parts corneous; dome-shaped appearance of cephalodistal portion not nearly so well marked as in first form male.

Allotypic Female.—The outstanding difference between the first form male and the female is in the chela which is never bearded within and always weaker (see measurements).

Annulus ventralis subrectangular in shape with a subtriangular projection on both lateral surfaces; deeply excavate cephalad with tuberculate prominences laterad; caudal margin slightly excavate; sinus originates dextrad of midventral line, bends sharply sinistrad, and runs laterad to cross the midventral line where it forms an arc and terminates just before it cuts the caudal margin.

Measurements.—Male (form I) Holotype: carapace, height 1.34, width 1.35, length 2.60 cm.; areola, width .19, length .87 cm.; rostrum, width .46, length .53 cm.; abdomen, length 3.0 cm.; right chela, length of inner margin of palm .88, width of palm .83, length of outer margin of hand 2.23, length of movable finger 1.14 cm. Female Allotype: carapace, height 1.41, width 1.45, length 2.80 cm.; areola, width .24, length .91 cm.; rostrum, width .53, length .60 cm.; abdomen, length 3.10 cm.; right chela, length of inner margin of palm .64, width of palm .73, length of outer margin of hand 1.78, length of movable finger 1.02 cm.

Type Locality.—A cypress pond and roadside ditch 9.4 miles north (State Highway 82) of Lake City, Columbia County, Florida. This locality is in the flatwoods area which extends westward and southwestward from the Okefenokee Swamp. Most of the specimens collected here were taken from simple burrows about one to one and one-half feet in depth and marked by low, poorly formed chimneys.

Disposition of Types.—The male holotype, the female allotype, and a second form male paratype are deposited in the United States National Museum. Of the remaining paratypes one male (form I), another male (form II), and one female are deposited in the Museum of Comparative Zoology; one male (form I), another male (form II), and a female in the University of Michigan Museum of Zoology; one male (form I), one male (form II), and a female in the Academy of Natural Sciences at Philadelphia. Fourteen males (form I), 18 males (form II), 56 females, four immature males, and four immature females are in my personal collection at the University of Florida.

Remarks.—My series of approximately 250 specimens does not represent a homogeneous population. Although all are probably conspecific the specimens from each locality tend to show small but distinguishable variations that enable one to recognize fairly accurately the region from which they came. In general some five variants may be recognized, and in the discussion that follows each is designated by the name of the town nearest to its place of capture.

One distinct variant occurs in the vicinity of Baxley, (Appling County) Georgia. The male is provided with hooks on the ischiopodites of only the third pair of pereiopods, but in one or two specimens there is a rudiment of a hook on the ischiopodite of the fourth. The areola is relatively broad with about three punctations in the narrowest part, and the rostrum is broader than in specimens from some of the other localities; the annulus ventralis bears no tubercles.

The males of the Jacksonville, Florida, specimens possess hooks on the ischiopodites of the third and fourth pereiopods. The areola is narrower and longer than in the Baxley specimens, and the annulus ventralis bears a group of three to five acute tubercles along the lateral margins.

In the males of the Waycross, Georgia, specimens hooks are present on the ischiopodites of both the third and fourth pereiopods. The rostrum as well as the areola is broader than that in the Jacksonville specimens, and the annulus ventralis, although similar to that of the Baxley females, has the lateral margins comparatively higher.

Like the Baxley males, those from Adel, (Cook County) Georgia, possess hooks on the ischiopodites of only the third pereiopods and can be distinguished from the Baxley specimens only by the structure of the first pleopod. The areolae are also similar, but the annuli ventralis of the Adel females bear an irregular group of three or four tubercles along each lateral surface.

The Lake City, Florida, specimens are most closely allied to those from Jacksonville but may be recognized by the difference in structure of the first pleopod. The surface of the annulus ventralis, though irregular, does not bear acute tubercles.

The specimens from Nassau County, Florida, are most like those from Jacksonville; those from Alachua, Columbia, and Union counties are for the most part similar to the Lake City specimens. The annuli ventralis in the three females from Flagler County are without tubercles, but until a first form male has been collected I hesitate to state what the relationships of the Flagler County specimens are.

Procambarus pubischelae probably has its closest affinities with *P. barbatus* and *P. escambiensis*. It is possible that *P. pubischelae* will prove to be a subspecies of *P. barbatus*, but present data indicate that they are geographically isolated in the region of the Altamaha River. On Georgia State Highway 38, between Hinesville and the river, I took specimens of *P. barbatus*. In this region flatwoods occur for a mile or two immediately north of the Altamaha, but the southern bank consists of a broad upland which extends southward almost to Jesup. About one-half mile north of Jesup *P. pubischelae* was collected. Other locality records strengthen the evidence that *P. barbatus* occurs only north of the Altamaha while *P. pubischelae* occurs only south of it.

Specimens Examined.—I have examined a total of 245 specimens of *Procambarus pubischelae*, from the following counties in Georgia and Florida: GEORGIA—Appling, Camden, Clinch, Colquitt, Cook, Lanier, Lowndes, Ware, and Wayne. FLORIDA—Alachua, Baker, Columbia, Duval, Flagler (?), Nassau, and Union.

SEASONAL DATA

	Jan.	Feb.	Mar.	Apr.	May	June	July	Aug.	Sept.	Oct.	Nov.	Dec.
♂ I					6	4		9		18	1	3
♂ II		3	5	3	1		1			19		
♀		2	10		12	4		2		51		2
♀ (eggs)				1	2					11		
♀ (young)										9		
♂ (immature)						1				6		8
♀ (immature)				11						5		17

Geographical and Ecological Distribution.—*Procambarus pubischelae* is probably confined to southeastern Georgia and northeastern Florida, where it is abundant in flatwoods situations. The northern boundary of its range seems to be in the vicinity of the Altamaha River, and while my collections in central Georgia are too inadequate to cite the western boundary in that state, this species is probably, with few exceptions, confined to the regions of the Coastal Terraces (specimens have been collected in Appling, Colquitt, Cook, Lowndes, Clinch, and Ware counties). In Florida the western boundary of the range lies in Columbia County, and probably follows the western limits of the flatwoods [shown in Harper's Generalized Soil Map of Florida (1925)] in Hamilton, Columbia, Bradford, and Alachua counties. The southern limit of the range seems to be approximately a line drawn from Lake Butler to Orange Heights, then due east to the Putnam County line. (No representatives of this species have been found in Putnam, Clay, or St. Johns counties, although several females, which if not *P. pubischelae* are very closely related to it, were collected 11 miles southeast of San Mateo, Flagler County.) The range probably extends eastward to the salt marshes and brackish inlets of the ocean in Georgia, while the eastern limit in Florida is still questionable.

Procambarus pubischelae was collected with *P. alleni* in Flagler County, with *P. seminolae* in Alachua, Baker, Duval, Union, and Nassau counties, with *P. fallax* in Alachua County, and with *P. paeninsulanus* in Baker County.

P. pubischelae is a characteristic secondary burrower. It is found abundantly (usually in groups) in the roadside ditches of the flatwoods. For the most part the burrows are extremely simple, consisting of a single tunnel which leads almost straight downward. Sometimes a burrow is slightly angulated or curved and terminates in a slight enlargement. The mouth is generally marked by a low, crudely formed chimney; occasionally, however, carefully built chimneys, in which the pellets are arranged in similar fashion to those of *P. advena*, are constructed. The burrows seldom exceed two and one-half feet in depth and have been observed in sandy clay soils, in black muck, in sandy mud, and in mottled red and blue clay.

I saw scores of open burrows scattered over the bottom of a shallow, temporary flatwoods pond west of Jacksonville shortly after a rainy season. Each burrow was about one and one-half feet deep and was marked by a small patch of yellow sandy clay around its opening. At the mouths of many of the burrows the crayfish could be seen with their chelae extended toward the opening, and, upon being disturbed, retreated into their holes. Occasionally, as I waded through the pond a crayfish would dart from a small clump of vegetation, scurry for a distance of one to four feet, and then disappear into one of the burrows. Some of these burrows had two openings, but all of them were otherwise very simple.

It is likely that most of the life of this species is spent in and about the mouth of the burrow, the crayfish leaving it only in search of food or in search of a mate. During the day I have seen this species in open water only once, and then only a few feet away from the mouth of a burrow. At night, however, specimens are often found in open water several feet from a burrow.

Occasionally a male and female have been found in the same burrow, and in each of these instances the male was in the first form. Generally there is an inch or two of water in the burrows, but I have often dissected burrows which, although moist, contained no water. In the latter cases the crayfish were apparently in good condition, not visibly affected by the lack of standing water.

Procambarus escambiensis, sp. nov.

Plate II, Figs. 11-15; Plate XIV; Maps 3, 4

Cambarus barbatus Faxon 1890 (in part), Proc. U. S. Nat. Mus. 12 (785): 621.

Diagnosis.—Rostrum without lateral spines; areola relatively broad with three to five punctations in narrowest part; male with hooks on ischiopodites of third and fourth pereiopods; palm of chela of first form male bearded within; postorbital ridges terminating cephalad without spines; no lateral spines present on carapace. First pleopod of male, form I, reaching coxopodite of second pereiopod and terminating in four distinct and corneous parts; mesial process long and spiniform (extending beyond the other terminal processes) and directed caudodistad; cephalic process small and subspiculiform; caudal process blade-like, compressed laterally; central projection also blade-like, but compressed obliquely cephalocaudad. Annulus ventralis subovate; longitudinal excavation along midventral line; high multituberculate ridges on either side of cephalic half, flattened caudad; sinus originates dextrad of midventral line about one-third of the total length of annulus from cephalic border; curves sharply laterad to cross midventral line and then gently caudad and finally dextrad to cut the caudal margin.

Holotypic Male, Form I.—Body subovate, compressed laterally. Abdomen narrower than thorax (1.25-1.54 cm. in widest parts respectively). Width and depth of carapace subequal in region of caudodorsal margin of cervical groove. Greatest width of carapace slightly caudad of caudodorsal margin of cervical groove.

Areola of moderate width (4.36 times longer than wide) with three punctations in narrowest part; sides parallel for a short distance in middle. Cephalic section of carapace about 2.36 times as long as areola (length of areola 29.7% of entire length of carapace).

Rostrum shallowly excavate above, reaching base of distal segment of peduncle of antennule; margins converging to tip; no lateral spines present. Upper surface of rostrum polished with a few scattered punctations. Marginal ridges high and sharp. Subrostral ridges well defined and evident in dorsal view for over half the length of the rostrum.

Postorbital ridge well defined, merging abruptly cephalad into the carapace without forming a tubercle or spine. Suborbital angle obtuse but prominent; branchiostegal spine well developed. No lateral spines or tubercles on sides of carapace. Surface of carapace with conspicuous punctations dorsad, granulate laterad.

Abdomen and thorax subequal in length.

Cephalic section of telson with two spines in each caudolateral corner.

Epistome subtrapezoidal in shape with a strong cephalomedian spine.

Antennules of usual form. A spine present on ventral side of basal segment; ventral surface conspicuously hirsute.

Antennae reaching caudad to fifth abdominal segment; antennal scale broad; broadest slightly distad of middle. Spine on outer margin strong.

Chela subovate, compressed dorsoventrally, broad and moderately long. Hand entirely tuberculate. Inner margin of palm densely bearded with plumose setae. A distinct ridge present on upper surface of both fingers. Fingers slightly gaping. Opposable margin of dactyl with a distinct excision at midlength, proximad of which are three dome-shaped tubercles, distad of which are six; interspersed between these are minute denticles. Lateral margin of dactyl with a proximal row of about eight subsquamous ciliated tubercles which terminates slightly distad of excision on opposable margin. Distad of this row of tubercles is a row of about seven setiferous punctations. Upper surface of dactyl with a submedian ridge flanked proximad by a group of several squamous tubercles and distad by scattered setiferous punctations along upper opposable surface and by a row of similar punctations laterad. Lower surface with a weak ridge flanked distad by a row of setiferous punctations and proximad by small tubercles mesiad and punctations laterad. Opposable margin of immovable finger with eight dome-shaped tubercles and a single large one projecting from lower surface at base of distal fifth. Interspersed among these are scattered minute denticles. Lateral margin with a poorly defined rounded ridge, most prominent near midlength of finger. Proximal third of upper surface with scattered squamous ciliated tubercles; distal portion with scattered setiferous punctations. Lower surface of immovable finger similar to that of dactyl.

Carpus of first pereiopod longer than wide (1.06-.80 cm.); shorter than inner margin of palm of chela (1.20 cm.); a distinct longitudinal groove above. Dorsomesial surfaces tuberculate, otherwise sparsely punctate. Six prominent tubercles on mesial surface and three on dorsal mesiodistal margin.

Merus of first pereiopod punctate laterad and mesiad. A group of scattered tubercles present on upper surface, more numerous distad. Lower surface with an irregular outer row of 18 spike-like tubereles and an inner row of 15.

Maxillipeds conspicuously heavily bearded.

Hooks present on ischiopodites of third and fourth pereiopods. Both pairs of hooks extending proximad over basiopodites. Bases of coxopodites of fourth and fifth pereiopods with strong outgrowths directed caudoventrad and cephaloventrad respectively. Outgrowth on fourth strongest but simple; outgrowth on fifth bituberculate.

First pleopod reaching second pereiopod when abdomen is flexed. Terminal portion of appendage with cephalodistal setae-bearing prominence rounded. Tip of pleopod ending in four distinct parts. Mesial process corneous, long, and spiniform, directed caudodistad and extending distad beyond the other terminal processes. Cephalic process which is small, corneous, and subspiculiform arises from mesial surface and directed distad. Caudal process corneous and blade-like, compressed laterally and lies at the caudolateral margin of the tip. Central projection also blade-like and corneous, is compressed cephalocaudad, and directed obliquely (cephalomesiad to cephalolaterad) across the center of the tip.

Male, Form II.—Differs from the male of the first form in the following respects: all spines and tubercles greatly reduced in size and number; palm of chelae not bearded; fingers not gaping, nor is there an excision on dactyl; all four terminals of pleopod evident, but none are corneous, and all blunt; central projection and caudal process indistinctly separated.

Allotypic Female.—Differs from the first form male in the following respects: palm of chela not bearded and weaker; no excision on dactyl.

Annulus ventralis subovate; longitudinal excavation along midventral line; high multituberculate ridges on either side of cephalic half; flattened caudad; sinus originates dextrad of midventral line about one-third of the total length of annulus from cephalic border; curves sharply laterad to cross midventral line and gently caudad and finally dextrad to cut caudal margin.

Measurements.—Male (form I) Holotype: carapace, height 1.64, width 1.54, length 3.23 cm.; areola, width .22, length .96 cm.; rostrum, width .48, length .60 cm.; abdomen, length 3.21 cm.; right chela, length of inner margin of palm 1.20, width of palm 1.10, length of outer margin of hand 2.82, length of movable finger 1.57 cm. Female Allotype: carapace, height 1.16, width 1.12, length 2.42 cm.; areola, width .17, length 1.73 cm.; rostrum, width .40, length .46 cm.; abdomen, length 2.60 cm.; right chela, length of inner margin of palm .62, width of palm .55, length of outer margin of hand 1.50, length of movable finger .81 cm.

Type Locality.—About 100 yards east of the Perdido River on U. S. Highway 90, Escambia County, Florida. All of my specimens of this species were taken from burrows in the type locality which is a small seepage area and drainage ditch. The soil consists of clay overlain by about one to two feet of black muck.

Disposition of Types.—The male holotype, the female allotype, and a second form male paratype are deposited in the United States National Museum. Of the remaining paratypes, one male (form I) and one female are deposited in the Museum of Comparative Zoology; one male (form I) and one female in the University of Michigan Museum of Zoology. One male (form I), three females, and one immature female are in my personal collection at the University of Florida.

Specimens at the Museum of Comparative Zoology, No. 3845, including one male, form I, three females, three immature males, and one immature female, are also designated paratypes of *P. escambiensis*.

Remarks.—Faxon (1890: 621), under the heading "Additional locality", listed the following for *C. barbatus*: "Escambia River at Flomaton, above Pensacola, Florida. D. S. Jordan, B. W. Evermann, and C. H. Bollman (M. C. Z.). One male, form I; five females, five young." In Faxon (1914: 414) these specimens are listed as "Paratypes" of *Cambarus barbatus*. I have examined these "Paratypes" and found that they belong to *Procambarus escambiensis*. This species is closely related to *Procambarus barbatus* and might easily be confused with it; however, a careful examination of the first pleopod of the male readily distinguishes *Procambarus escambiensis* from all other known species. In an attempt to get this crayfish in the Escambia River, I have collected at several points along its course in Florida and Alabama, but have been unable to find it either in the river or burrowing in its flood plains. Instead I found another closely related form, *Procambarus shermani*. It would thus seem questionable as to whether the Flomaton, Alabama, labels on the specimens of the species referred to *barbatus* by Faxon represent an actual collection in this vicinity.

In its pubescent chelae and other general body structures, *P. escambiensis* most closely resembles *P. pubischelae*. The structure of the first pleopod of the male, however, indicates that its closest relative is *P. econfinae*.

As was pointed out in the discussion of the distribution of the *barbatus* group, *P. escambiensis* probably came into this region from the north and west by a route that paralleled the movement of the *econfinae* and *shermani* stocks east of it.

Specimens Examined.—I have examined a total of 11 specimens of *Pro-*

cambarus escambiensis, from Escambia County, Florida, and all of my specimens were taken during the month of April [4 ♂ ♂ (form 1), 1 ♂ (form II), 6 ♀ ♀, and 1 ♀ (immature)]. I have also seen and made camera lucida drawings of the specimens, mentioned above, in the Museum of Comparative Zoology.

Geographical and Ecological Distribution.—*Procambarus escambiensis* is known from only two localities, on the western periphery of the range of the *barbatus* group. In my collection I have specimens from a single locality, a low wet place at the Perdido River on U. S. Highway 90. The specimens at the Museum of Comparative Zoology were said to have been taken from the Escambia River. Even if these specimens were taken from the Escambia River region, which I questioned above, I do not believe that they came from the river itself but probably from an overflow pool or nearby marshy area. All of the species of this subgroup are burrowers, and although occasionally I have seen them in open water I have never observed them in a large or swift stream.

Cambarus byersi was taken from burrows along with *P. escambiensis* in the type locality.

P. escambiensis is probably a secondary burrowing species. The simple burrows were one and one-half to two feet deep in clay and muck; the water table was from six inches to a foot below the surface. The chimneys over the burrows were somewhat demolished by recent rains but were probably poorly constructed.

Procambarus econfinae, sp. nov.

Plate II, Figs. 16-20; Plate XV; Maps 3, 4

Diagnosis.—Rostrum broadly lanceolate; areola broad and short with space for about six punctations in narrowest part; however, sparsely punctate; male with hooks on ischiopodites of third and fourth pereiopods; palm of chela of first form male not bearded along inner margin; postorbital ridges terminating cephalad without tubercles or spines; no lateral spines present on carapace. First pleopod of male, form I, reaching coxopodite of third pereiopod and terminating in four distinct, corneous parts; mesial process subspiculiform, directed distad, and extending distad of the other terminal processes. Cephalic process small and directed cephalodistad but hidden in lateral view by the terminal setae. Caudal process compressed laterally, broadly rounded, and forming a high ridge across the distal caudolateral margin of the appendage. Central projection small and blade-like, and directed cephalolaterad. Cephalodistal setae-bearing prominence somewhat dome-like. Annulus ventralis subovate; sinus originates on midventral line slightly caudad of cephalic margin; turns gently sinistrad forming a broad arc back to the midventral line along which it proceeds almost to the caudal margin of the annulus.

Holotypic Male, Form I.—Body subovate, compressed laterally. Abdomen longer than thorax (2.45-2.30 cm.). Height of carapace greater than width (1.17-1.15 cm.). Greatest width of carapace slightly caudad of caudodorsal margin of cervical groove.

Areola broad and short (2.52 times longer than wide); sparsely punctate, however, with three or four in narrowest part; sides parallel in middle. Cephalic section of carapace about 2.7 times as long as areola (length of areola 27.4% of entire length of carapace).

Rostrum broad-lanceolate, flattened above with low lateral ridges; margins converging to tip; no lateral spines present. Upper surface of rostrum punctate. Subrostral ridges evident in dorsal view for only a short distance near base.

Postorbital ridges well defined terminating cephalad without spines or tubercles. Suborbital angle moderately well developed, rounded; branchiostegal spine small. No lateral spines on sides of carapace. Surface of carapace punctate dorsad, granulate laterad.

Abdomen longer than thorax (2.45-2.3 cm.).

Cephalic section of telson with two spines in each caudolateral corner.

Epistome broadly oval with slightly undulant cephalolateral margins.

Antennules of the usual form, not bearded. A spine present on ventral surface of basal segment.

Antennae reaching caudal margin of first abdominal segment (apparently injured but would probably extend no farther back than the third abdominal segment); antennal scale broad; broadest slightly distad of middle. Spine on outer margin small.

Chela broadly ovate, compressed, heavy, entirely tuberculate, and with slender fingers. Inner margin of palm not bearded but bearing an irregular row of about 10 tubercles. A well defined ridge present on upper surface of both fingers. Fingers slightly gaping. Opposable margin of dactyl with 16 rounded tubercles between which are a few scattered denticles. Lateral surface of dactyl ridge-like, bearing five tubercles on basal half and about 10 setiferous punctations on distal half. A few scattered punctations on upper proximolateral surface. Upper surface with a distinct submedian ridge flanked by setiferous punctations. Lower surface of finger with a longitudinal ridge near outer margin; otherwise with setiferous punctations. Opposable margin of immovable finger with 12 rounded tubercles, and in addition a larger one extending from lower surface at base of distal fourth. Minute denticles interspersed among these tubercles. Outer margin of immovable finger with a distinct excavate ridge bearing setiferous punctations. Upper and lower surfaces with submedian ridges flanked proximally by tubercles and distally by setiferous punctations.

Carpus of first pereiopod longer than broad (.63-.51 cm.); shorter than inner margin of palm of chela (.81 cm.); well defined longitudinal groove above. Dorsomesial and mesial surfaces studded with tubercles. Only four prominent ones on mesial surface. Upper mesiodistal surface with three prominent tubercles.

Merus punctate mesiad and laterad with a few tubercles on upper distal surface. Lower surface with an irregular outer row of about 12 tubercles and a more regular inner one of about 14.

Hooks present on ischiopodites of third and fourth pereiopods. Hooks on fourth small. Bases of coxopodites of fourth pereiopods with no prominent outgrowths; corresponding position on fifth pereiopod with a small recurved plate-like hook.

First pleopod reaching base of third pereiopod. Cephalodistal setae-bearing prominence somewhat dome-like. Mesial process long, subspiculiform, corneous, and directed distad and extending distad of the other terminal processes. Cephalic process long and corneous, directed cephalodistad but hidden in lateral view by the terminal setae. Caudal process compressed laterally, broadly rounded and corneous, and forming a high ridge across the distal caudolateral margin of the appendage. Central projection, blade-like, corneous, and directed cephalolaterad.

Male, Form II.—The description of the first form male will suffice for the male of the second form if it is kept in mind that the tubercles are smaller but more spiniform. First pleopod with all of the processes evident and occupying the same relative position; however, none of them are corneous, and all are much reduced in size.

Allotypic Female.—Likewise the description of the first form male is applicable to the female except in the general form of the chelae, which are smaller and apparently weaker.

Annulus ventralis subovate. Sinus originates on midventral line slightly caudad of cephalic margin; turns gently sinistrad forming a broad arc back to the midventral line along which it proceeds almost to the caudal margin of the annulus.

Measurements.—Male (form I) Holotype: carapace, height 1.17, width 1.15, length 2.30 cm.; areola, width .25, length .63 cm.; rostrum, width .50, length .43 cm.; abdomen, length .25 cm.; right chela, length of inner margin of palm .81, width of palm .81, length of outer margin of hand 2.04, length of movable finger 1.17 cm. Female Allotype: carapace, height 1.18, width 1.14, length 2.40 cm.; areola, width .21, length .65 cm.; rostrum, width .44, length .49 cm.; abdomen, length 2.65 cm.; left chela, length of inner margin of palm .62, width of palm .64, length of outer margin of hand 1.50, length of movable finger .90 cm.

Type Locality.—Flatwoods in the northern part of Panama City, Bay County, Florida. Here a large colony of this species occupies a small wet area lying between the railroad and U. S. Highway 231 near the northern city limits. This area is subject to flooding in wet weather; however, at the time I collected here the water table was about one foot beneath the surface. All of my specimens were dug from burrows.

Disposition of Types.—The male holotype, the female allotype, and a second form male paratype are deposited in the United States National Museum. Of the remaining paratypes one male (form I) and one female are deposited in the Museum of Comparative Zoology; one male (form I) and one female in the University of Michigan Museum of Zoology. Twelve males (form I), one male (form II), and 28 females are in my personal collection at the University of Florida.

Remarks.—*Procambarus econfinae* is an extremely localized species apparently occupying a total area of less than 200 square miles. My series of specimens shows little variation. Its nearest relatives are *P. escambiensis*, *P. apalachicolae*, and *P. latipleurum*. The first pleopod very closely approximates that of *P. escambiensis* while the palm of the chela is naked as in the other species just mentioned.

Specimens Examined.—I have examined a total of 48 specimens, from two localities in the vicinity of Panama City, Bay County, Florida [northern part of city: June 1938—10 ♂ ♂ (form I), 1 ♂ (form II), 13 ♀ ♀, 3 ♀ ♀ with young, 1 ♀ with eggs; the same locality: April 1938—2 ♂ ♂ (form I), 1 ♂ (form II), 2 ♀ ♀ ; 7.9 miles northeast of Panama City on U. S. Highway 231: June 1938—3 ♂ ♂ (form I), 8 ♀ ♀, 2 ♀ ♀ with eggs, 2 ♀ ♀ with young].

Geographical and Ecological Distribution.—*Procambarus econfinae* is confined to a small area of the flatwoods on the small peninsula on which Panama City is located. The northern boundary of its range is marked by North Bay and by a band of well drained soils of the Blanton-Norfolk group; the western boundary is marked by a portion of the bay and soils of a similar type along with the Lakewood-St. Lucie group on the southwest, while to the south East Bay acts as a barrier. Only to the east is there no apparent barrier delimiting the range of this species unless the brackish nature of the small streams which cut across this area, and the marshy Wetappo Intercoastal

Canal prevent an eastward migration. A discussion of the possible origin of this species was taken up in the discussion of the *barbatus* group.

P. econfinae was collected along with *P. pycnogonopodus* in Bay County.

Almost certainly this species, like its relatives, is a secondary burrower. Both of the localities from which it was taken are in the flatwoods where the water table is only a foot or so below the surface, and standing water is present in them during the rainy season. Here the crayfish construct simple burrows consisting of a single passage downward, which never branches more than once. The burrows are from one to three feet deep and are marked by low, somewhat crude chimneys. In several instances I found a male and a female in the same burrow. In one of these burrows a female carrying young was accompanied by a male; in another burrow having two branches a male and female were taken from one branch and a lone female from the other. In still another branched burrow a male was found in one pocket and a female in the other. This is the only species I have collected in which a female carrying young was accompanied by a male.

Procambarus latipleurum, sp. nov.

Plate III, Figs. 21-25; Plate XVI; Maps 3, 4

Diagnosis.—Rostrum narrow and acute lanceolate with or without lateral spines; if spines are present they are small and close to the tip; areola relatively broad with about six punctations in narrowest part; males with hooks on ischiopodites of third and fourth pereiopods, hooks on fourth bulbiform; palm of chela of first form male not bearded within; chela slender and weak with no (except sometimes one on immovable finger) well developed tubercles on opposable surfaces of fingers; chela slender and weak; postorbital ridges terminating cephalad in a small tubercle or spine; no lateral spines present on carapace. First pleopod of male, form I, reaching coxopodite of third pereiopod and terminating in four distinct parts; mesial process corneous, extremely long, spiculiform, and directed distolaterad, distinctly curved; cephalic process small and corneous, forming a hook which somewhat overhangs the central projection; caudal process somewhat truncate but having a corneous caudodistal edge; a distinct depression on lateral surface of main body of appendage at base of this process; central projection small and corneous, compressed cephalocaudad and directed cephalodistad, somewhat overhung by the cephalic process; cephalodistal setae-bearing surface rounded; pleura of abdomen conspicuously broad. Annulus ventralis almost square in outline, lateral sides rounded; small acute tubercles present on cephalolateral margins; sinus originates on cephalic margin along midventral line and runs caudad slightly dextrad of the midventral line to middle of annulus where it makes a sharp turn across the midventral line forming an arc and terminates just before it reaches the caudal margin, slightly sinistrad of the midventral line.

Holotypic Male, Form I.—Body subovate, compressed laterally. Abdomen and thorax subequal in width. Height of carapace slightly greater than width (1.55-1.42

cm.). Greatest width of carapace about midway between caudodorsal margin of cervical groove and caudal margin of carapace.

Areola of moderate width (4.65 times longer than wide) with six punctations in narrowest part; sides parallel in middle. Cephalic section of carapace about 1.96 times as long as areola (length of areola 33.8% of entire length of carapace).

Rostrum acute-lanceolate, flattened above with rather high marginal ridges; margins converging to tip; no lateral spines present. Upper surface of rostrum punctate. Subrostral ridges well defined, barely visible in dorsal view almost to tip.

Postorbital ridge well defined, terminating cephalad in a small tubercle. Suborbital angle obtuse and rounded; branchiostegal spine well developed. No lateral spines on sides of carapace. Surface of carapace punctate dorsad, weakly granulate laterad.

Abdomen much longer than thorax (3.42-2.75 cm.).

Cephalic section of telson with two spines in each caudolateral corner.

Epistome fan-shaped with a faint cephalomedian projection.

Antennules of usual form, but heavily bearded. A spine present on ventral surface of basal segment.

Antennae reaching caudal margin of fourth abdominal segment; antennal scale broad; broadest in middle. Spine on outer margin moderately strong.

Chela subovate, compressed, long and slender, entirely tuberculate; tubercles thick. Inner margin of palm with two rows of about nine tubercles each. A very poorly defined submedian ridge on upper surface of each finger. Fingers not gaping. Opposable margin of dactyl with an upper and lower row of very small tubercles bordering a wide band of minute denticles covering the entire margin. Upper surface of both fingers bearing setiferous punctations except near base where there are a few scattered tubercles. Median ridges poorly defined. Lateral and ventral surfaces with a few tubercles at base, otherwise with setiferous punctations. The outstanding feature of the chela of this species is the absence of well developed tubercles on opposable margins of fingers.

Carpus of first pereiopod longer than wide (.72-.46 cm.); shorter than inner margin of palm of chela (.82 cm.); faint indication of a longitudinal groove above. Dorsomesial, mesial, and ventral surfaces studded with tubercles, otherwise with setiferous punctations. No acute tubercles present on ventral surface and about four prominent ones on mesial surface.

Merus of first pleopod punctate laterad, tuberculate dorsad and mesiodistad. A group of tubercles present on dorsodistal surface. Lower surface with a very irregular outer row of about 15 tubercles and an inner row of about 16.

Hooks present on ischiopodites of third and fourth pereiopods, hooks on fourth pereiopod bulbiform. Coxopodites of fourth and fifth pereiopods with distinct outgrowths; those on the fourth heavy and directed ventrolaterad; those on the fifth small and slender and directed ventrad.

First pleopod reaching base of third pereiopod when the abdomen is flexed. Terminal portion of appendage with the cephalodistal setae-bearing surface rounded. Mesial process corneous, extremely long, spiculiform, and directed distolaterad; distinctly curved. Cephalic process small and corneous, forming a hook which somewhat overhangs the central projection. Caudal process somewhat truncate but having a corneous caudodistal edge; a distinct depression on lateral surface of main body of appendage at base of this process. Central projection small and corneous, compressed cephalocaudad and directed cephalodistad, somewhat overhung by the cephalic process.

Pleura of abdomen distinctly long and broad.

Male, Form II.—Differs from the first form male in the following respects: small lateral spines are present on rostrum near tip; tubercles reduced in size; postorbital ridges terminate cephalad in small spines; epistome with less regular margins. First pleopod of male with scarcely a distinction between any of the terminal processes except the mesial which is large and claw-like, curved cephalodistad, and directed slightly laterad.

Allotypic Female.—Differs from the first form male in the following respects: epistome with an acute strong cephalomedian projection; chela broader and shorter.

Annulus ventralis almost square in outline; lateral sides rounded; small acute tubercles present on cephalolateral margins; sinus originates on cephalic margin along midventral line and runs caudad slightly dextrad of the midventral line to middle of annulus where it makes a sharp turn across the midventral line forming an arc and terminates just before reaching the caudal margin, slightly sinistrad of the midventral line.

Measurements.—Male (form I) Holotype: carapace, height 1.55, width 1.42, length 2.75 cm.; areola, width .20, length .93 cm.; rostrum, width .40, length .58 cm.; abdomen, length 3.43 cm.; right chela, length of inner margin of palm .82, width of palm .64, length of outer margin of hand 2.06, length of movable finger 1.0 cm. Female Allotype: carapace, height 1.58, width 1.45, length 2.82 cm.; abdomen, length 3.41 cm.; areola, width .21, length .87 cm.; rostrum, width .42, length .70 cm.; right chela, length of inner margin of palm .55, width of palm .61, length of outer margin of hand 1.62, length of movable finger .97 cm.

Type Locality.—Roadside excavation and intermittent stream in flatwoods, 5.8 miles west of Weewahitchka on State Highway 52, Gulf County, Florida. Most of my specimens were taken with a seine from the excavation in which the water was from one to three feet deep. A few specimens were dug from simple burrows, never with more than two branches, nor more than two feet deep. The chimneys of most of these burrows were neatly formed.

Disposition of Types.—The male holotype, the female allotype, and a second form male paratype are deposited in the United States National Museum. Of the remaining paratypes one male (form II) and a female are deposited in the Museum of Comparative Zoology; one male (form II) and one female in the University of Michigan Museum of Zoology. One male (form I), seven males (form II), two females, three immature males, and two immature females are in my personal collection at the University of Florida.

Remarks.—All of my specimens were taken within a radius of 13 miles of Weewahitchka.

This species has its closest affinities with *P. apalachicolae, P. rathbunae, P. pubischelae,* and *P. barbatus.* It is indeed surprising to find this local form in the Apalachicola Flatwoods without apparent barriers to separate it from *P. apalachicolae.* The break between these two species occurs somewhere in the 25 mile wide extent between Weewahitchka and Port St. Joe. Westward it is probably limited by the well drained area of Blanton-Norfolk soils.

Specimens Examined.—I have examined a total of 54 specimens of *Procambarus latipleurum,* from Gulf County, Florida, collected during the months of April, October, and November. [April—21 ♂ ♂ (immature) and 11 ♀ ♀ (immature); October—2 ♂ ♂ (form I), 9 ♂ ♂ (form II), 2 ♀ ♀, 3 ♂ ♂ (immature), and 2 ♀ ♀ (immature); November—1 ♂ (form II) and 3 ♀ ♀.]

Geographical and Ecological Distribution.—*Procambarus latipleurum* has been taken only in the flatwoods around Weewahitchka, and judging from the fact that a closely related species has been collected in abundance both south and west of its range and that ample barriers bar its migration northward and eastward, it is probably confined to a small area in this region.

This species was collected along with *P. pycnogonopodus*, *P. pygmaeus*, and *P. kilbyi*.

This species is most likely a secondary burrower and is apparently confined to flatwoods conditions where it occupies temporary bodies of water. Its burrows are simple, never having more than two branches, usually consisting of a single vertical tunnel which penetrates the water table within two feet below the surface of the ground. The chimneys vary in height from one to four inches and are neatly formed, having vertical rather than sloping outer walls.

Procambarus apalachicolae, sp. nov.
Plate III, Figs. 26-30; Plate XVII; Maps 3, 4

Diagnosis.—Rostrum broad, short, and without lateral spines; areola relatively broad with about four punctations in narrowest part; male with hooks on ischiopodites of third and fourth pereiopods; palm of chela of first form male not bearded within; chela heavy with relatively short fingers; postorbital ridges terminating cephalad without lateral spines or tubercles; no lateral spines present on carapace. First pleopod of first form male reaching coxopodite of third pereiopod and terminating in four distinct parts; mesial process acute with a corneous tip (extending caudodistad beyond the other terminal processes); cephalic process small, corneous, blade-like, and truncate distad; caudal process large, non-corneous, and thumb-like; central projection corneous, small, subtriangular, compressed cephalocaudad, and situated obliquely cephalomesiad to caudolaterad; cephalodistal setae-bearing surface rounded. Annulus ventralis with cephalic and caudal margins concave, lateral margins convex; sinus originates on cephalic surface on midventral line, curves sinistrad forming a wide arc and terminates just cephalad of midcaudal margin; a few small tubercles on each cephalolateral surface.

Holotypic Male, Form I.—Body subovate, compressed laterally. Abdomen narrower than thorax (.89-1.0 cm. in widest parts respectively). Height of carapace slightly greater than width in region of caudodorsal margin of cervical groove (1.05-1.0 cm.). Greatest width of carapace slightly caudad of caudodorsal margin of cervical groove.

Areola moderately broad (3.7 times longer than wide) with only two punctations in narrowest part (however, there is room for four or five, were they spaced as they are elsewhere in the areola); sides not parallel. Cephalic section of carapace about 2.48 times as long as areola (length of areola about 28.7% of the entire length of carapace).

Rostrum flattened above, almost reaching base of distal segment of peduncle of antennule; margins converging to tip; no lateral spines present. Upper surface of rostrum punctate. Marginal ridges strongly developed and terminate before reaching tip of rostrum. Subrostral ridges prominent and evident in dorsal view to tip of rostrum.

Postorbital ridges well defined, terminating cephalad without forming a tubercle or spine. Suborbital angle obtuse; branchiostegal spine small. No lateral spines or tubercles on sides of carapace. Surface of carapace punctate dorsad, weakly granulate laterad.

Abdomen longer than thorax (2.23-1.95 cm.).

Cephalic section of telson with four spines in the sinistral and three in the dextral caudolateral corners.

Epistome broadly ovate with a small cephalomedian projection; margins slightly undulate.

Antennule of usual form with a small spine present on ventral side of basal segment.

Antennae reaching caudad to third abdominal segment; antennal scale moderately broad; broadest in middle. Spine on outer margin well developed.

Chela subovate, compressed dorsoventrally, moderately heavy, and entirely tuberculate. Inner margin of palm not bearded but bearing an irregular row of about 11 tubercles. A distinct ridge present on upper surface of both fingers. Fingers not gaping. Opposable margin of dactyl with six rounded tubercles on proximal three-fifths and several rows of minute denticles distad of these; denticles are also interspersed between tubercles. Upper distal margin with a few minute tubercles above the denticles. Lateral surface of dactyl with a row of eight tubercles and several setiferous punctations. Upper and lower surfaces each with a submedian longitudinal ridge flanked by scattered tubercles proximad and setiferous punctations distad. Opposable margin of immovable finger with ten rounded tubercles and a larger one projecting from lower margin at base of distal fifth; interspersed between these tubercles is a large number of minute denticles; lateral surface with an excavate ridge bearing setiferous punctations; upper and lower surfaces with a submedian ridge flanked proximally by scattered tubercles, distally by setiferous punctations.

Carpus of first pereiopod longer than wide (.63-.43 cm.); a poorly defined longitudinal groove above. Dorsomesial and mesial surfaces tuberculate, otherwise punctate. Mesial surface with several scattered tubercles. Cephalomesial and dorsomesial margins with a continuous row of about seven tubercles.

Merus of first pereiopod punctate on lateral and proximomesial surfaces. Upper surface with scattered tubercles over entire length. Lower surface with an irregular outer row of about 15 tubercles and an inner row of about 17.

Hooks present on ischiopodites of third and fourth pereiopods. Coxopodites of fourth and fifth pereiopods with outgrowths. Those on the fourth heavy, subtriangular, and directed ventrolaterad; those on the fifth thin, somewhat rounded, and similarly directed.

First pleopod reaching coxopodite of third pereiopod when the abdomen is flexed. Terminal portion of appendage with cephalodistal setae-bearing prominence rounded. Mesial process acute, with corneous tip, and extending distad beyond the other terminal apices. Cephalic process, small, slender, truncate and corneous, arises from the mesial surface and is directed distad. Caudal process thumb-like, non-corneous, and directed caudodistad. Central projection corneous, small, triangular, compressed cephalocaudad, and extends obliquely, cephalomesiad to caudolaterad.

Male, Form II.—Differs from the first form male in the following respects: tubercles fewer in number; many, however, more spiniform; subrostral ridges evident in dorsal aspect for only a short distance from base and at tip of rostrum. First pleopod with all terminals evident, but much reduced and non-corneous.

Allotypic Female.—Differs from the first form male in the following respects: ridges of rostrum extending to tip; subrostral ridges evident in dorsal aspect along basal half of rostrum; chela proportionately smaller; tubercles somewhat reduced in size.

Annulus ventralis with cephalic and caudal margins concave, lateral margins convex. Sinus originates on cephalic surface on midventral line; curves sinistrad forming a wide arc and terminates just cephalad of midcaudal margin. A few small tubercles on each cephalolateral surface.

Measurements.—Male (form I) Holotype: carapace, height 1.05, width 1.0, length 1.95 cm.; areola, width .15, length .56 cm.; rostrum, width .31, length .62 cm.; abdomen, length 2.23 cm.; right chela, length of inner margin of palm .62, width of palm .54, length of outer margin of hand 1.57, length of movable finger .86 cm. Female Allotype: carapace, height 1.30, width 1.21, length 2.34 cm.; areola, width .23, length .69 cm.; rostrum, width .38, length .50 cm.; abdomen, length 1.95 cm.; right chela, length of inner margin of palm .53, width of palm .63, length of outer margin of hand 1.42, length of movable finger .85 cm.

Type Locality.—Roadside ditch in wire grass flatwoods, 11.1 miles west of Beacon Hill, U. S. Highway 98, Bay County, Florida. All of the specimens were dug from simple burrows in a sandy, mucky soil.

Disposition of Types.—The male holotype, the female allotype, and a second form male paratype are deposited in the United States National Museum. Of the remaining paratypes one male (form I) and one female are deposited in the Museum of Comparative Zoology; one male (form I) and one female in the University of Michigan Museum of Zoology. Eight males (form I), 12 females, and one immature male are in my personal collection at the University of Florida.

Remarks.—*Procambarus apalachicolae* occupies a rather peculiar range when the range of its relatives, *P. econfinae* and *P. latipleurum*, are taken into consideration. While the ranges of these three species are not entirely separated by apparent physiographic barriers, I find no instance of intergradation.

The nearest relatives of this species are almost certainly *P. latipleurum* and *P. rathbunae*, and it is possible that *P. apalachicolae* will prove to be a subspecies of the latter.

Specimens Examined.—I have examined a total of 85 specimens, from the following counties in Florida: Bay, Franklin, Gulf, and Walton. All of these localities were in the flatwoods along the coast from the Apalachicola River westward to the mouth of Choctawhatchee Bay.

SEASONAL DATA

	Jan.	Feb.	Mar.	Apr.	May	June	July	Aug.	Sept.	Oct.	Nov.	Dec.
♂ I				6	12	9				1		
♂ II				3		1						
♀				12	13	6						
♀ (eggs)					1	2						
♀ (young)						1						
♂ (immature)				7	1							
♀ (immature)				14								

Geographical and Ecological Distribution.—As was pointed out above *Procambarus apalachicolae*, *P. econfinae*, and *P. latipleurum* present an intricate pattern of distribution in the Apalachicola flatwoods. The ranges of the last two have already been discussed, but it is necessary to recall the limits of their ranges in connection with that of *P. apalachicolae*. The three species together occupy practically all of the available flatwoods between the Apalachicola River on the east and the Choctawhatchee River and Bay on the west. *P. latipleurum* occupies the northeastern part of the area, and there is no known barrier to the south to separate it from *P. apalachicolae*. In collecting along State Highway 6 between Weewahitchka and Port St. Joe the most northern locality for *P. apalachicolae* was 5.6 miles north of Port St. Joe, while the most southern record for *P. latipleurum* was 12.6 miles south of Weewahitchka; hence, along this road there is a zone of six miles from which neither species has been taken. The frequently inundated marshes bordering the Wetappo Intercoastal Canal occur within this area and may be a limiting factor in the distribution of these two species. Thus, on the basis of the material at hand it seems as though the northern limit of the range of *P. apalachicolae* is along the Wetappo Intercoastal Canal and the East Bay. It is also consistent with the evidence at hand that this same marsh bordered canal to the west in Bay County may also mark the limits of the ranges of *apalachicolae* and *econfinae*.

Apparently the range of *P. apalachicolae* is completely broken by St. Andrews and North bays and a strip of well drained soils to the north of North Bay. In western Bay County and in Walton County *apalachicolae* seems to be scattered throughout the flatwoods as far as 12 miles east of Destin. The southern boundary of the range is apparently marked by the salt marshes and dunes along the coast.

P. apalachicolae was taken along with *P. pycnogonopodus*, *P. rogersi rogersi*, and *P. rogersi* (intergrades) in Bay County, and with *P. rogersi* (intergrades) in Franklin County as well as in Gulf County.

Like the other members of the *barbatus* group, *P. apalachicolae* is a secondary burrower. Typically, it is a flatwoods species living in depressions in which water is present during some part of the year. When the depression becomes dry the crayfish construct simple burrows with small chimneys. These chimneys show a large variety of patterns; some are apparently very carefully constructed, with straight vertical walls; others are low and have the appearance of small mud piles that show little resemblance to a typical chimney. In most of the stations where this species has been collected the water table was about one foot below the surface, but in several instances it was as deep as three feet. Burrows have been found in a wide variety of soils: in sands, clays, and muck. In most instances wire grass has been present in the vicinity where this species has been taken, and in at least one locality gallberries and palmettos were abundant.

In a small roadside ditch having a vertical bank and in which water was standing several specimens of this species had dug burrows into the bank, with openings both above and below the water table; several openings were as high as three feet above the existing water level, and most of these had tunnels which were about one foot below the level of the water.

In two instances specimens of *apalachicolae* were taken from standing water; several specimens were dipped from a bed of *Juncus repens* in a small swamp pool, and others were taken in the isolated pools of a drying swamp stream.

Procambarus rathbunae (Hobbs)

Plate III, Figs. 31-35; Maps 3, 4

Cambarus rathbunae Hobbs 1940a, Proc. U. S. Nat. Mus. 89 (3097): 414-418, Fig. 21; also 387, 389.

Procambarus rathbunae Hobbs, Amer. Mid. Nat. (in press-c).

Diagnosis.—Rostrum without lateral spines; areola relatively broad with three or four punctations in narrowest part; male with hooks on ischiopodites of third pereiopods only; palm of chela not bearded along inner margin; postorbital ridges terminating cephalad without spines; no lateral spines present on carapace. First pleopod of male reaching coxopodite of second pereiopod and terminating in four distinct parts; mesial process extends distad beyond the rest of the terminal elements in a caudodistal direction; cephalic process small and corneous and arises from mesial surface of the appendage and is directed distad; caudal process, making up the caudolateral surface of the tip, forms a distinct corneous ridge which is directed cephalodistad; central projection small, subtriangular, corneous, and somewhat compressed cephalocaudad with a small rounded plate laterad. Cephalodistal setae-bearing prominence sloping. Annulus ventralis broadly subovate with the smaller ends in the longitudinal axis; however, truncate both cephalad and caudad; excavate along midventral line with a transverse ridge on either side at midlength which is abrupt caudad but gradually sloping cephalad; cephalic portion on either side of excavation tuberculate; sinus originates about one-third of the length of the annulus from the cephalic border, slightly dextrad of the midventral line, runs caudosinistrad barely crossing the midventral line where it abruptly turns and forms a straight line to the midcaudal margin.

Remarks.—*Procambarus rathbunae* is most nearly related to *P. apalachicolae*; in fact a few specimens of the latter species from the southwestern part of Bay County and the southeastern part of Walton County are structurally intermediate between the two species. It seems doubtful that they represent intergrades, however, because continuous barriers of salt water and sand hills appear to completely isolate the two. The possibility of intergradation cannot be dismissed, however; a thorough search should be made in Walton, Okaloosa, and Washington counties before it can be finally determined whether the specimens from the Destin Region are really a variant of *Procambarus apalachicolae*, or whether they are *apalachicolae* x *rathbunae*. *P. pubischelae* and *P. escambiensis* are also near relatives of this species.

Specimens Examined.—I have examined a total of 27 specimens of *Procambarus rathbunae*, from Okaloosa and Holmes counties, Florida.

Seasonal Data.—I have collected this species during the months of April, May, and October. In April 1938 two males (form I), five males (form II), and nine females were taken from burrows and a ditch at Milligan. In May 1942 one male (form I) and one female were collected eight miles west of Ponce de Leon on U. S. Highway 90. In October 1941 two males (form II), four females (one of which was carrying young), and three immature females were taken 3.6 miles south of Crestview on State Highway 54.

Geographical and Ecological Distribution.—It is likely that this species is widespread in the flood plains of the Yellow River and seepage areas and flatwoods along its tributaries in Santa Rosa and Okaloosa counties. It is possible that it will be found in the Blackwater River drainage, for there are continuous low marshes between the mouths of the two rivers which may possibly form a highway for this species. I am unwilling at present to make any statement concerning its possible distribution in the Choctawhatchee drainage, for I have it from only one locality, while *P. hubbelli,* another member of the *barbatus* group occupying similar habitats, has been found to be widespread in this river system in Florida.

P. okaloosae was captured in the type locality of *P. rathbunae* both in the aquatic vegetation of a roadside ditch and from burrows adjacent to those of *P. rathbunae.*

All except one of the specimens of *P. rathbunae* taken at Milligan were dug from simple burrows in the roadside ditches. This single specimen was caught in a dip net dragged along the ditch. The burrows ranged in depth from six inches to a foot below the surface of the ground. Several of the burrows were open, though most of them were marked by closed chimneys. The soil is a sand-clay mixture, and the ground is covered with a thick mat of grass. Some sections of the ditch held water, but at this time most of it was dry.

Near Crestview and Ponce de Leon *P. rathbunae* was taken from a sloping seepage area along the margins of a bay around which wire grass and pitcher plants were common. The burrows were extremely complex, closely resembling those of the *rogersi* subspecies. They were from one to three feet in depth and branched several times. Some of them had several chimneys, most of which were poorly formed, and in several places where excavations had been made along the hillside open burrows were numerous in their walls. In no place in the area was the water table more than a foot below the surface.

While collecting at Milligan I encountered an open burrow in which a male was either attempting to enter or else being driven out by the female. I could see both crayfish since the opening of the burrow was unusually large. The male seemed to be attempting to enter, and the female, facing him, was apparently blocking the burrow.

On the basis of my limited observations the habits of *P. rathbunae* are similar to those of the other better known species of the *barbatus* group.

THE KILBYI, SHERMANI, AND HUBBELLI SUBGROUPS

The three remaining species of the *barbatus* group are so distinct from the members of the *barbatus* subgroup and from one another that each necessitates a separate subgroup to indicate its isolated position.

THE SHERMANI SUBGROUP

A separate *shermani* subgroup is necessitated to receive the very disjunct *Procambarus shermani.*

Procambarus shermani[8], sp. nov.

Plate III, Figs. 36-40; Plate XVIII; Maps 3, 4

Diagnosis.—Rostrum without lateral spines; areola relatively narrow (narrower than in any other species in the *barbatus* group) with two punctations in the narrowest part; male with hooks on ischiopodites of third and fourth pereiopods; palm of chela of first form male usually bearded; postorbital ridges terminating cephalad in small tubercles; a lateral spine present on carapace. First pleopod of first form male reaching coxopodite of second pereiopod and terminating in four distinct parts; mesial process acute and corneous; cephalic process and central projection blade-like, corneous, and compressed laterally; both directed caudodistad; caudal process subspatulate and directed laterodistad with a small corneous ridge arising on the lateral surface and flanking it along its caudal margin; no cephalodistal setae-bearing prominence; setae borne on the slightly curved cephalodistal margin. Annulus ventralis subovate with the greatest length in the longitudinal axis; deep submedian furrow flanked on either side by high multituberculate wall, deepest portion of furrow in center of annulus; sinus originates in bottom of furrow slightly dextrad of midventral line, runs gently caudosinistrad, crossing midventral line where it turns caudodextrad terminating just before reaching caudal margin, slightly dextrad of midventral line.

Holotypic Male, Form I.—Body subovate, compressed laterally. Abdomen narrower than thorax (1.25-1.40 cm. in widest parts respectively). Width and depth of carapace subequal in region of caudodorsal margin of cervical groove. Greatest width of carapace slightly caudad of caudodorsal margin of cervical groove.

Areola relatively narrow (12.9 times longer than wide) with two punctations in narrowest part; sides parallel in middle. Cephalic section of carapace about 1.89 times as long as areola (length of areola 34.6% of entire length of carapace).

Rostrum acute lanceolate, flattened above but with conspicuously high lateral ridges, reaching midlength of distal segment of peduncle of antennule; margins converging to tip; no lateral spines present. Upper surface of rostrum with shallow mosaic furrows. Subrostral ridges weakly developed and not evident in dorsal view.

Postorbital ridge very prominent and terminating cephalad in a tubercle. Suborbital angle obtuse, rounded; branchiostegal spine well developed. A single lateral spine present on sides of carapace, flanked by several small tubercles. Surface of carapace punctate dorsad, strongly granulate laterad.

Abdomen and thorax subequal in length.

Cephalic section of telson with three spines in the right and four in the left caudolateral corners.

Epistome broadly oval with a small, acute cephalomedian projection.

Antennule of usual form. A spine present on ventral side of basal segment.

Antennae reaching caudad to third abdominal segment; antennal scale of moderate length, widest slightly distad of middle. Spine on outer margin strong.

---

[8]This species is named for Professor H. B. Sherman, Department of Biology, University of Florida, who has frequently aided me in my work on the Florida crayfishes, not only in the loan of personal equipment but also in invaluable criticisms and suggestions.

Chela subovate, flattened dorsoventrally, of moderate length and width. Hand entirely tuberculate. Inner margin of palm bearded with plumose setae. A distinct ridge present on upper surface of both fingers. Fingers not gaping. Opposable margin of dactyl convex, with a faint indication of an excision at end of proximal third, proximad of which are five dome-shaped tubercles, distad of which are 14, all of which are interspersed with minute denticles. Lateral margin of dactyl with a row of eight prominent tubercles. Upper surface of dactyl with a submedian ridge at the base of which is a group of small squamous tubercles, more numerous laterad; a single row of setiferous punctations along lateral margin of ridge, likewise a single row along upper surface of opposable margin. Lower surface with a prominent, though not distinctly defined, ridge flanked proximad by a few scattered tubercles and distad by several rows of setiferous punctations. Opposable margin of immovable finger with 20 dome-shaped tubercles and a single large, more acute one projecting from lower surface at base of distal fourth. Interspersed among these are scattered minute denticles. Lateral margin of immovable finger with a ridge flanked proximally by ciliated squamous tubercles and distally by setiferous punctations. Lower surface of immovable finger with a prominent ridge, and ventromesial surface distinctly excavate. A row of setiferous punctations runs the entire length of the finger just below opposable margin, and a few tubercles are present on inner side of ridge at base.

Carpus of first pereiopod longer than wide (.73-.59 cm.); shorter than inner margin of palm of chela (.89 cm.); a distinct longitudinal groove above. Dorsomesial and mesial surfaces tuberculate. Eight prominent tubercles on mesial surface, twelve on dorsomesial surface, and an additional very strong one on distal dorsomesial margin. Ventral and ventromesial surfaces each with a single spine on distal margin.

Merus of first pereiopod punctate laterad and proximomesiad, otherwise tuberculate. Two very prominent spines on upper surface near distal end; one prominent spine on same surface at distal margin, below which are two prominent tubercles. Lower surface with an irregular outer row of about 20 tubercles and an inner row of about 16 spike-like tubercles.

Maxillipeds bearded.

Hooks present on ischiopodites of third and fourth pereiopods. Bases of coxopodites of fourth and fifth pereiopods with strong outgrowths directed caudoventrad and cephaloventrad respectively.

First pleopod reaching base of second pereiopod when abdomen is flexed; with no cephalodistal setae-bearing prominence; setae borne on the slightly curved cephalodistal margin. Tip ending in four distinct parts. Mesial process acute, corneous, and does not extend beyond the other terminal processes. The other three elements are subequal in length; cephalic process blade-like and corneous, as is the central projection; both are compressed laterally and directed slightly caudodistad. Caudal process subspatulate and corneous, directed distolaterad with a small corneous ridge arising on the lateral surface and flanking it around its caudal margin. The cephalic and caudal processes and the central projection together form a thin corneous blade across the lateral tip of the appendage.

Male, Form II.—Differs from the male of the first form in the following respects: all tubercles greatly reduced in size and number; however, all spiny portions accentuated; palm of chela lightly bearded; postorbital ridges bearing spines. First pleopod with truncate terminals; three small terminals representing the cephalic and caudal processes and the central projection; mesial process shortened and less acute.

Allotypic Female.—Differs from the first form male in the following respects: palm of chela not bearded and weaker; cephalic margin of epistome broken by several tubercles.

Annulus ventralis subovate with the greatest length in the longitudinal axis; deep submedian furrow flanked on either side by a high multituberculate wall; deepest portion of furrow in center of annulus; sinus originates in bottom of furrow slightly dextrad of midventral line; runs gently caudosinistrad, crossing midventral line where it turns caudodextrad terminating just before reaching caudal margin, slightly dextrad of midventral line.

Measurements.—Male (form I) Holotype: carapace, height 1.40, width 1.40, length 2.93 cm.; areola, width .08, length 1.03 cm.; rostrum, width .50, length .40 cm.; abdomen, length .30 cm.; right chela, length of inner margin of palm .89, width of palm .78, length of outer margin of hand 2.30, length of movable finger 1.27 cm. Female Allotype: carapace, height 1.90, width 1.84, length 4.02 cm.; areola, width .06, length 1.48 cm.; rostrum, width .60, length .90 cm.; abdomen, length 4.00 cm.; right chela, length of inner margin of palm .82, width of palm .80, length of outer margin of hand 2.46, length of movable finger 1.47 cm.

Type Locality.—About 12 miles southwest of Jay, Santa Rosa County, Florida, in the Escambia River swamp along McCaskill's Mill Creek. The crayfish were taken from flood plain pools among dead leaves and other debris. Burrows were numerous in the bottoms of these pools, and at night the crayfish could be seen at the mouths of them. I collected here both at night and during the day, using a headlight and dip net at night and a seine during the day. A larger catch was obtained during the daylight hours. At the time of my visit to the swamp the river was flooded, and even during my brief stay of less than 24 hours the water had risen so that practically all of the isolated pools were connected and a swift current was flowing through many of them.

Disposition of Types.—The male holotype, the female allotype, and a second form male paratype are deposited in the United States National Museum. Of the remaining paratypes one male (form I) and one female are deposited in the Museum of Comparative Zoology; one male (form I) and one female in the University of Michigan Museum of Zoology. Two males (form I), one male (form II), four females, 22 immature males, and 21 immature females are in my personal collection at the University of Florida.

Remarks.—The first specimens of this species that I saw were caught by Mr. Ray Boles in swamp pools in the river bottom a few miles south of the type locality. One second form male and three immature females were taken along with a series of *P. okaloosae* and *P. bivittatus*. It was not until April 1942 that I succeeded in getting a series of this species.

P. shermani somewhat bridges the gap between the *barbatus* section and the section of *blandingii*. In the structure of the chelae and body conformation, with the exception of the presence of lateral spines on the carapace, it fits nicely into the *barbatus* section. The pleopod, however, is intermediate between that found in the two sections. The cephalic process has moved forward and extends from the cephalic margin rather than the mesial margin. Also the caudal process is corneous and very similar to that found in the *blandingii* subspecies and *hayi*.

Procambarus shermani like the rest of the species of the *barbatus* group probably invaded this region from the north, paralleling the migration of *escambiensis* in the Perdido River system.

The closest relatives of this species seem to be *Procambarus escambiensis, P. latipleurum, P. rathbunae, P. econfinae, P. apalachicolae*, and *P. pubischelae*.

Specimens Examined.—I have examined a total of 57 specimens of *Procambarus shermani*, from Santa Rosa County, Florida, and all of them were taken within three miles of the type locality. Four specimens were collected in May 1941: 1 ♂ (form II) and 3 ♀ ♀ (immature). In April 1942 the following were taken: 5 ♂ ♂ (form I), 1 ♂ (form II), 7 ♀ ♀, 22 ♂ ♂ (immature), and 18 ♀ ♀ (immature).

Geographical and Ecological Distribution.—*Procambarus shermani* has been taken only from the Escambia River swamp. The fact that a closely allied species occurs in the Perdido River drainage system and another in the Yellow River drainage system, both of which occupy the same general type of habitat as that from which *shermani* was taken, would seem to indicate that the latter is confined to the Escambia River drainage.

P. shermani was collected along with *P. blandingii acutus, C. diogenes,* and *C. species incertis* in Santa Rosa County.

Practically nothing is known of the habits of this species. It is apparently confined to swamp pools or sluggish water, for it was not taken in McCaskill's Mill Creek which is a clear sand-bottom, moderately flowing stream, but was relatively abundant in the small pools near its margins. It apparently burrows in these pools, but even in the middle of the day specimens may be taken from leaf mats and other debris.

As was pointed out under the heading "Type Locality," the Escambia River is subject to considerable fluctuation in water level. In October 1941 I was able to walk all over the swamp, even down to the river's edge, but in April 1942 the water was so high that I could not even get into the swamp at the place I visited before, and it was only with considerable difficulty that collecting was carried out in the swamp pools farther upstream along McCaskill's Mill Creek. The creek itself was completely obliterated in the mass of water in the swamp. Collecting could be successfully carried out in only a few isolated pools.

THE KILBYI SUBGROUP

A separate *kilbyi* subgroup is necessitated to receive the very disjunct *Procambarus kilbyi*.

Procambarus kilbyi (Hobbs)
Plate IV, Figs. 41-45; Maps 3, 4

Cambarus kilbyi Hobbs 1940a, Proc. U. S. Nat. Mus. 89 (3097): 410-414, fig. 20; also 387-389, 418.
Cambarus kilbyi Hobbs (in press-a).
Procambarus kilbyi Hobbs (in press-c); Hobbs (in press-d); Hobbs (in press-b).

Diagnosis.—Rostrum without lateral spines; areola relatively broad; male with hooks on ischiopodites of third and fourth pereiopods; palm of chela not bearded along inner margin; postorbital ridges terminating cephalad without spines; no lateral spines present on carapace. First pleopod of first form male reaching coxopodite of third pereiopod and terminating in four distinct parts; mesial process, the largest of the four, corneous, subspatulate, and bent near midlength at about a 50 degree angle to the main shaft of the appendage; cephalic process arises from the mesial surface of the appendage and is slender, subspiniform, and truncate; caudal process somewhat rounded, compressed laterally, and extends in a ventrocaudal direction from the caudolateral margin of the appendage; central projection small, corneous, triangular, and compressed cephalocaudad; cephalodistal setae-bearing prominence knob-like. Annulus ventralis subovate; sinus originates on cephalic border slightly sinistrad of midventral line, curves gently dextrad of midventral line, then somewhat more sharply sinistrad across the midventral line, and finally to the midcaudal margin.

Remarks.—Although *P. kilbyi* is decidedly variable and occupies a relatively large area, unlike several of the other species of the *barbatus* group, it is not on the basis of the material at hand divisible into local variants. While the general appearance of any specimen enables one to tell with a reasonable degree of certainty from what locality it was taken, I do not have sufficient material to determine the extent of these variants in a single local population.

There is a tendency for the outline of the annulus ventralis to become more angulate in specimens from the southeastern portion of the range. The two extremes in the shape of this structure are found in specimens from Calhoun and Levy counties. In Calhoun County the annulus is subovate with the angles rounded, while in the Levy County females it is decidedly angulate. The interesting fact is that there is a somewhat gradual transition between the two types (although not without exceptions) in the intervening region.

While the first pleopod of the male shows considerable variation, all of these differences seem to be individual ones.

Specimens Examined.—I have examined a total of 305 specimens of *Procambarus kilbyi*, from the following counties in Florida: Calhoun, Franklin, Gulf, Jackson, Jefferson, Lafayette, Leon, Levy, Liberty, Madison, Taylor, and Wakulla.

SEASONAL DATA

	Jan.	Feb.	Mar.	Apr.	May	June	July	Aug.	Sept.	Oct.	Nov.	Dec.
♂ I		5	4	10	9	3				2	2	
♂ II		5		15	4	7				7	5	6
♀		5	9	11	16	5		2	1	6	7	5
♀ (eggs)					3	1						
♀ (young)					4	2						
♂ (immature)		8	4	37	1				1	5	2	4
♀ (immature)		11	7	46	4				2	9	4	4

Geographical and Ecological Distribution.—*Procambarus kilbyi* is a Florida endemic and seems to be confined to the coastal flatwoods from Levy County westward into Calhoun and Gulf counties. Exactly what, if anything, has checked the westward migration of this species has not been determined;

there is no apparent barrier to the west of the most western known record. The Apalachicola River probably formed a temporary barrier, but now that *kilbyi* has established itself west of the river I can see nothing which can restrict its farther westward migration unless *P. apalachicolae* proves too much of a competitor. It is possible that *P. kilbyi* will migrate farther into Bay, Gulf, and Walton counties in the future. The eastern boundary of its range is limited by the high, dry soils which extend from Hamilton County southward through Marion County into the center of the state. The southern limit of the range is not definitely known, though the record from Levy County represents the most southern locality (about 6 miles southwest of Bronson [State Highway 19]). It is very possible that this species extends farther south along the coast; certainly, however, it goes no farther south than Hernando County. The southwestern limit of the range is apparently marked by the dunes and salt marshes along the gulf.

P. kilbyi has been collected with *P. rogersi campestris* in Leon and Wakulla counties, with *P. rogersi ochlocknensis* in Liberty County, with *P. rogersi rogersi* in Calhoun County, with *P. rogersi* intergrades in Franklin County, with *P. pycnogonopodus* in Gulf, Jackson, and Calhoun counties, with *P. spiculifer* in Gulf and Jackson counties, with *P. apalachicolae* and *P. latipleurum* in Gulf County, with *P. leonensis* in Jefferson, Liberty, Wakulla, and Madison counties, with *P. pygmaeus* in Liberty County, and with *P. paeninsulanus* in Madison and Taylor counties.

Although *P. kilbyi* probably should be classified as a secondary burrowing species it is not unusual to find it in open water, in temporary puddles, or even in small temporary streams. The type specimens were taken from a small, temporary creek by means of a dip net which was pushed through the vegetation.

In Leon County about 12 miles south of Tallahassee the chimneys of *P. rogersi campestris* were common around a typical sour-gum and ti-ti bay which is very low and subject to flooding in wet weather. These burrows had chimneys which ranged from two to six inches in height. Near by at the edge of a small bay in sandy soil and in a region where the water table was about one to three feet below the surface, I found burrows which were less complex than the majority of the burrows I had been digging, and in them were the first specimens of *P. kilbyi* that I had seen. Wire-grass, palmetto, gallberry, and small ti-ti were characteristic of the flora. Pitcher plants and sundews were also common.

In Gulf County about 13 miles south of Weewahitchka specimens were taken from burrows which were marked by neatly formed chimneys about five inches high. In the ditch close by smaller specimens of *P. kilbyi, P. latipleurum,* and *P. pycnogonopodus* were taken.

In Taylor County about six miles north of Perry I waded through a roadside excavation which was about a foot deep in water. Scattered drowned burrows were observed and dug, but no crayfish were found in them. However, by wading through the grass which covered the bottom I startled an occasional crayfish which would dart some ten or twelve feet ahead, where if I was careful to keep from disturbing it a second time it could be bagged with a dip net. *P. kilbyi* is probably the fastest swimmer of any of the Florida crayfishes, and because it blends so perfectly with the environment, is one of the hardest species to catch in open water.

In a small temporary stream near Blountstown I scooped up hundreds of specimens of very young *kilbyi*, and upon uprooting some of the vegetation found that the bottom of the stream was riddled with scores of open burrows.

Many other collecting records give similar ecological data. All point to the fact that *P. kilbyi* is a coastal flatwoods species and apparently is just as well suited to open water as to the burrowing habit. All of the stations in which this species has been taken from open water are subject to becoming dry, and this makes it necessary that the crayfish occupy burrows for a part of the year. All of the females with eggs or young that I have seen have been taken from burrows.

The Hubbelli Subgroup

A separate *hubbelli* subgroup is necessitated to receive the very disjunct *P. hubbelli*.

Procambarus hubbelli (Hobbs)

Plate IV, Figs. 46-50; Maps 3, 4

Cambarus hubbelli Hobbs 1938b (nomen nudum), Proc. Fla. Acad. Sci. 2: 90.
Cambarus hubbelli Hobbs 1940a: 406-410, fig. 19; also 387, 389, 414, 418.
Procambarus hubbelli Hobbs (in press-c).

Diagnosis.—Rostrum without lateral spines; areola relatively broad with three punctations in narrowest part; male with hooks on ischiopodites of third pereiopods only; palm of chela of first form male usually bearded along inner margin; postorbital ridges terminating cephalad without spines; no lateral spines present on carapace. First pleopod of first form male reaching coxopodite of third pereiopod and terminating in four distinct parts; mesial process spiniform and extends in a distal caudomesial direction; cephalic process consists of a small, corneous, triangular structure which somewhat shields the central projection cephalad; caudal process large, corneous, fanlike, and situated along the distal caudolateral surface; central projection small, corneous, laterally flattened, and lying somewhat beneath the cephalic process; cephalodistal setae-bearing prominence knob-like. Annulus ventralis subovate with the greatest length in the transverse axis; sinus originates on the cephalic margin slightly dextrad of the midventral line, curves rather abruptly sinistrad to cross the midventral line where it curves gently caudodextrad and terminates just before reaching the midcaudal margin.

Remarks.—None of the mostly small variations shown by my specimens indicates any correlation with any definite locality. The mesial processes of the first pleopods of several specimens from the southeastern part of Walton County are directed more mesiad than in specimens from other localities, but even there specimens are found in which the first pleopod is almost typical.

Procambarus hubbelli is probably the most disjunct species of the *barbatus* group. The first pleopod of the male gives no clue to its relationships except that it indicates rather distant affinities with the other species of the *barbatus* group. It seems most closely related to *P. escambiensis*, and this relationship is indicated by both morphological and geographical evidence. Both of these species have barbate chelae and show several other morphological similarities, and the proximity of the range of *P. escambiensis* is consistent with the morphological evidence.

Specimens Examined.—I have examined more than 300 specimens of *Procambarus hubbelli*, collected from Holmes, Jackson, Walton, and Washington counties, Florida.

SEASONAL DATA

	Jan.	Feb.	Mar.	Apr.	May	June	July	Aug.	Sept.	Oct.	Nov.	Dec.
♂ I			1	1	16	12						
♂ II			1	2	82	3						
♀			7	4	102	17						1
♀ (eggs)					1							
♀ (young)												
♂ (immature)					18							
♀ (immature)					44							

Geographical and Ecological Distribution.—*Procambarus hubbelli* is known from the western portion of Jackson County, from scattered localities in the southern part of Holmes County, from the southeastern part of Walton County, and from a few scattered localities in the northern and southwestern parts of Washington County. It is probable that this species is confined to the Choctawhatchee River drainage where it inhabits the flatwoods adjoining this river and Holmes Creek. Undoubtedly, this species is locked in, so to speak, by the sand ridges that run north and south in Walton County and northeast to southwest in the southeastern part of Washington County. It thus appears that the only highway for further active migration is along the coast in Walton and Bay counties.

Procambarus hubbelli was collected with *P. pycnogonopodus* in Walton, Washington, and Holmes counties, and with *P. paeninsulanus* in Holmes and Washington counties.

Like other members of the *barbatus* group, *P. hubbelli* is a secondary burrowing, flatwoods species. Although most of my specimens were taken from burrows, some were found to occur in sluggish creeks and temporary ponds in the flatwoods. They are common in roadside ditches and have been taken from burrows in low areas along creeks where wire-grasses, bog-inhabiting orchids, and pitcher plants are characteristic elements of the flora. The burrows are generally simple, and have only one or two main passages, which are almost invariably vertical. They have been found in almost pure clay, in coarse sand, and often in black mud. The chimneys are usually crudely constructed and have the appearance of a small pile of mud. An occasional chimney, however, seems to have been carefully built. I have seen the bottom of a small temporary pond in Holmes County literally riddled with burrows over an area of more than 100 square feet; here many of the crayfish were crouched in the mouths of the burrows, and upon being disturbed darted backward into their holes. Occasionally I have taken specimens from a submerged mud and leaf drift in the quiet reaches of a stream.

It is common to find a first form male and a female occupying the same burrow. In my collection there is only one female with eggs, and she was the only occupant of the burrow from which she was taken.

I have no reason to believe that this species has habits widely different from those of the other members of the *barbatus* group, so what has already been said concerning the habits of *P. pubischelae, escambiensis, kilbyi,* and *rathbunae* is in general applicable to *P. hubbelli*.

The Alleni Group

A separate *alleni* group is necessitated to receive the very disjunct *Procambarus alleni*.

Procambarus alleni (Faxon)
Plate IV, Figs. 51-55; Map 5

Cambarus alleni Faxon 1884, Proc. Amer. Acad. Arts and Sci. 20: 110-112, also 138.

Cambarus alleni Faxon 1885a: 18, 19, 35-36, 158, 173, Pl. I, fig. 1, Pl. VIII, figs. 2, 2'; Faxon 1890: 619-620, 621; Lönnberg 1894a: 125; Lönnberg 1894b: 3, 10; Faxon 1898: 646; Ortmann 1902: 277; Harris 1903: 58, 70, 97, 137, 143, 152, 166; Ortmann 1905a: 100, 102, 105, 127, 129; Ortmann 1905b: 403; Ortmann 1906: 18, 19; Faxon 1914: 370, 371, 415; Creaser 1931b: 4; Creaser 1934: 4.

Procambarus alleni Hobbs (in press-c).

Diagnosis.—Rostrum with or without lateral spines; areola relatively narrow (7-14 times as long as broad); hooks on ischiopodites of third and fourth pereiopods, those on fourth bituberculate; palm of chela not bearded along inner margin but bears from 9-11 tubercles; postorbital ridges terminating cephalad with or without spines; one lateral spine usually present on sides of carapace. First pleopod of male, form I, reaching coxopodite of third pereiopod and terminating in four parts; mesial process corneous, subspiculiform, S-shaped, and extends distad beyond the rest of the appendage; cephalic process arising from mesial side of appendage, is corneous and blade-like and its distal tip lies proximad of the other terminal apices; central projection a small, corneous blade, borne along mesial side of caudal process; caudal process large and finger-like, extends distad from caudolateral margin; cephalodistal setae-bearing surface forming an obtuse angle. Annulus ventralis subovate with a lateral wing-like projection on either side; sinus originates on midcephalic margin, runs caudad for two-thirds the length of the annulus and then bends sinistrad, making a hairpin turn to the midventral line where it curves caudad to cut the midcaudal margin of the annulus.

Remarks.—*Procambarus alleni* has been recorded as one of the several species in which the rostrum, postorbital ridges, and the cervical groove are without lateral spines. Most of the specimens in the museums, no doubt, exhibit these features, but in my collection which includes specimens from localities over a large portion of the range, all variations, from a complete lack of spiny parts to a condition equally as spiny as found in *P. fallax*, are represented. Because of this some of the females of the two species are almost indistinguishable.

P. alleni exhibits numerous minute variations, and although most of them are not radical changes, practically every structure generally used for a taxonomic character shows some degree of variation. Some of these are correlated with the habitat of the crayfish while others have no discernible correlations either with habitat or with locality.

Some local, probably inbred, populations are recognizable on a rather circumscribed combination of several of these variations, and although collections from other regions may exhibit the same list of variations, here they will rarely be shown in the phenotype of any one individual.

In general specimens of *P. alleni* from a stream or other permanent body of water are more spiny than are those collected from burrows or from pools that are definitely temporary.

Previous authors in discussing the relationships of *P. alleni* have assigned it to a wide variety of different groups, but at least one opinion has been shared by all of them—that it is an aberrant species. Although I do not think *alleni* is so aberrant as some of these authors have contended, it is geographically far removed from *P. simulans* which I believe to be its nearest relative. Ortmann (1905a: 100, 102) erected a separate group for *P. evermanni*, *P. barbatus*, *P. wiegmanni*, and *P. alleni*, selecting *alleni* as the type of the group. I agree that *alleni* and *barbatus* are allied; however, *P. evermanni* and *P. wiegmanni* are not correctly placed with the other two species. In view of the following similarities between *alleni* and *simulans* I do not think they should be referred to separate sections, but on the basis of the first pleopod I have placed them in separate groups. In both species the areola is relatively narrow, and the rostrum in many of the specimens of *alleni* is devoid of spines, as is the case in *simulans*. The general shape of the carapace is similar, and while the chelae are different, there are certain resemblances in them. The annulus ventralis of the female of *alleni* is definitely of the *simulans* type, and finally, the first pleopod of the male of *alleni* is closer to that of *simulans* than to that of any other member of the *barbatus* section. As would be expected the first pleopods of the second form males of the two species agree in detail to a much greater extent than do those of the first form males. Faxon has emphasized that in *simulans* only the third pair of pereiopods bears hooks, while in *alleni* hooks occur on both the third and fourth pereiopods. I do not think that this distinction need be given a great deal of weight, for it has been noted that the number of walking legs with hooks is variable—even within certain single species. It is noteworthy that *simulans*, *gracilis*, and *alleni* occupy respectively the western, the north-central, and the southeastern extremities of the range of the genus.[9]

The *alleni* stock probably migrated into Florida from the northwest shortly after the closing of the Suwannee Straits, to be followed at a later date by an influx of other species, *fallax*, *paeninsulanus*, *pubischelae*, etc.

The apparent geographical-ecological relationship of *P. alleni* to *P. fallax* is well marked outside of the range of *alleni*. The more ubiquitous *fallax* is found in habitats very similar to those occupied by *alleni* within the latter's own range. Where their ranges overlap both species have occasionally been taken from the same station, but generally *fallax* appears to have taken sole possession of the streams and the more permanent bodies of water. Since *fallax* occupies the area between the northwestern and northeastern extremities of the range of *alleni*, it seems at least possible that some competition is going on between these two species, and that *fallax* is the more successful competitor. In the Withlacoochee River, which cuts across the range of *alleni*, *fallax* seems to be the only crayfish inhabitant, and this is also true for the Hillsborough and Aliphia rivers and several streams in Hardee and Indian River counties, well within the range of *alleni*.

Specimens Examined.—I have examined approximately 620 specimens, from the following counties in Florida: Brevard, Broward, Charlotte, Citrus, Collier, Dade, Flagler, Glades, Hendry, Hillsborough, Indian River, Lake, Lee, Levy, Manatee, Martin, Monroe, Okeechobee, Orange, Osceola, Palm Beach, Pasco, Pinellas, Polk, Putnam, Sarasota, Seminole, St. Johns, St. Lucie, Sumter, and Volusia.

SEASONAL DATA

	Jan.	Feb.	Mar.	Apr.	May	June	July	Aug.	Sept.	Oct.	Nov.	Dec.
♂ I	8	9	5	38	3	1	7	2	19	18	29	10
♂ II	6	5	3	19	4		2	4	29	8	17	2
♀	17	8	10	51	8	1	14	11	47	53	39	24
♀ (eggs)				4	4					1		1
♀ (young)					1							
♂ (immature)	4			9			7	9		16	5	
♀ (immature)		2	1	5		1	2	5	3	12		

[9] The only other species of the *simulans* group, *hagenianus*, formerly reported from South Carolina (type locality cited as near Charleston) has not been taken from South Carolina since that time. Mr. G. Robert Lunz of the Charleston Museum has collected rather diligently for this species and has never succeeded in obtaining a specimen from South Carolina, and he doubts that the types came from that state. It has been definitely recorded from Alabama and Mississippi.

Geographical and Ecological Distribution.—The extent of the range of *P. alleni* probably exceeds that of any other species of crayfish in Florida. The species is abundant in certain localities east of the St. Johns River and has been collected at intervals along the east coast to Big Pine Key. The sinuous northern limit of the range of this species may be approximated by a line drawn along the St. Johns River from its mouth to northern Seminole County, thence westward to Bushnell in Sumter County, and northwestward through Lecanto and Citronelle in Citrus County, and northward through Rosewood in Levy County, and again northwestward to the mouth of the Suwannee River. The range probably extends eastward to the dunes and brackish waters along the Atlantic while the southern and western boundaries are marked by similar situations along the gulf.

Throughout the northern extent of its range *P. alleni* is associated with other species, and in at least one instance, it occurs in the same habitat with *paeninsulanus* at the boundaries of their ranges. A series of sand ridges cuts across State Highway 13 between the sixth and eighth mile posts north of Cedar Keys. In a collection made between the two most western ridges (about six miles north of Cedar Keys) all of the specimens were *P. alleni;* likewise a collection between the next ridges to the east, 6.3 miles northeast of Cedar Keys, gave only *P. alleni.* Between a still more eastward pair of dunes, 6.7 miles northeast of Cedar Keys, both *P. alleni* and *P. paeninsulanus* were taken. This is the most eastern record for *P. alleni* in Levy County. Within one-half mile to the east of this locality the sand dunes give way to low flatwoods that extend eastward to Bronson. This flatwoods region has been frequently collected, and although *paeninsulanus* was abundant, no specimens of *alleni* were found.

Faxon (1914) records *P. alleni* from Lake Butler, near Tarpon Springs, in Hillsborough County, and from ponds near Tampa. This species is abundant in the roadside ditches throughout the northern part of its range, and the cypress ponds in St. Johns County are thickly populated by it.

Cypress ponds seem to be a favorable habitat for this species. Often in wading through the shallow margins of these ponds which usually have a bottom of black silty mud, one observes the small mounds of lighter soil which mark the openings of the crayfish burrows. If one approaches one of these mounds cautiously the crayfish may be seen at the mouth of the burrow with its chelae extended. Some individuals dig their burrows beneath felled logs, and by moving submerged logs one often uncovers several burrows. (Occasionally I have seen *Amphiuma means* and *Siren intermedia* in small tunnels under such logs, and it is possible that these tunnels were constructed by crayfish.) I have seen *alleni* hiding in the submerged vegetation of roadside ditches; usually, however, if disturbed, at least some of the crayfish rapidly retreated into open burrows in the bottom of the ditch.

Although this species has been taken in flowing water, the streams seemed to be either temporary or very sluggish. The preferred habitat of *alleni* appears to be somewhat temporary, still bodies of water in a region where the soil will support a burrow.

The burrows of *alleni* are simple, vertical, or slanting, and range in depth from one to three feet. The chimneys appear as mounds of small pellets of mud or clay, and only occasionally does one have a definitely cylindrical structure.

About seven miles south of Dunnellon, Citrus County, I found burrows in pure sand, and when I attempted to dig the animals out many were lost because the sand sloughed off the walls of the burrows faster than I could dig it out. The water table in this case was only a few inches below the surface, and just how these burrows remained open I cannot understand. *Alleni* is the common flatwood species throughout the extent of its range and is the only species I have seen from the everglades.

THE ADVENA SECTION

Diagnosis.—The cephalodistal surface of the first pleopod of the first form male never terminates in a ridge or a knob-like prominence but in a corneous, reduced cephalic process, or if the cephalic process is absent, then the cephalodistal surface is almost flush with the centrocephalic process of the central projection. The mesial process is slender, spiniform or blade-like, and generally directed distad; the central projection is decidedly the most conspicuous terminal element, and is either laterally compressed or directed across the cephalodistal tip of the appendage. The caudal process is present as a large bump or thumb-like process. The rostrum is broad and short and without lateral spines; the areola is narrow or obliterated; the male has hooks on the ischiopodites of the third, or the third and fourth pereiopods; the chelae are compressed and bear a cristiform row of tubercles along the inner margin of the palm.

The *advena* section includes six species and subspecies (*advena, geodytes, pygmaeus, rogersi rogersi, rogersi ochlocknensis,* and *rogersi campestris*).

P. advena is with little doubt the most primitive member of the section. The ancestral stock probably migrated from the northwest. The easternmost branch of this stock then gave rise to the present day *advena* while the western branch advanced along the Ochlocknee River to the Apalachicola flatwoods where it gave rise to the *rogersi* group. It was along this latter path that *pygmaeus* (a member of the *advena* group) later moved into the same region. A portion of the *advena* group probably diverged to migrate southward into Florida along the eastern side of Trail Ridge. Part of this stock continued along the St. Johns River, and another turned southwestward into the region of Alachua County. Certainly *P. advena* has its closest affinities with *P. geodytes* and *P. pygmaeus*.

The *advena* section has been divided into two groups: the *advena* group and the *rogersi* group.

THE ADVENA GROUP

Diagnosis.—The mesial process of the first pleopod of the male is well developed, spiniform, or slightly compressed; the cephalic process is either

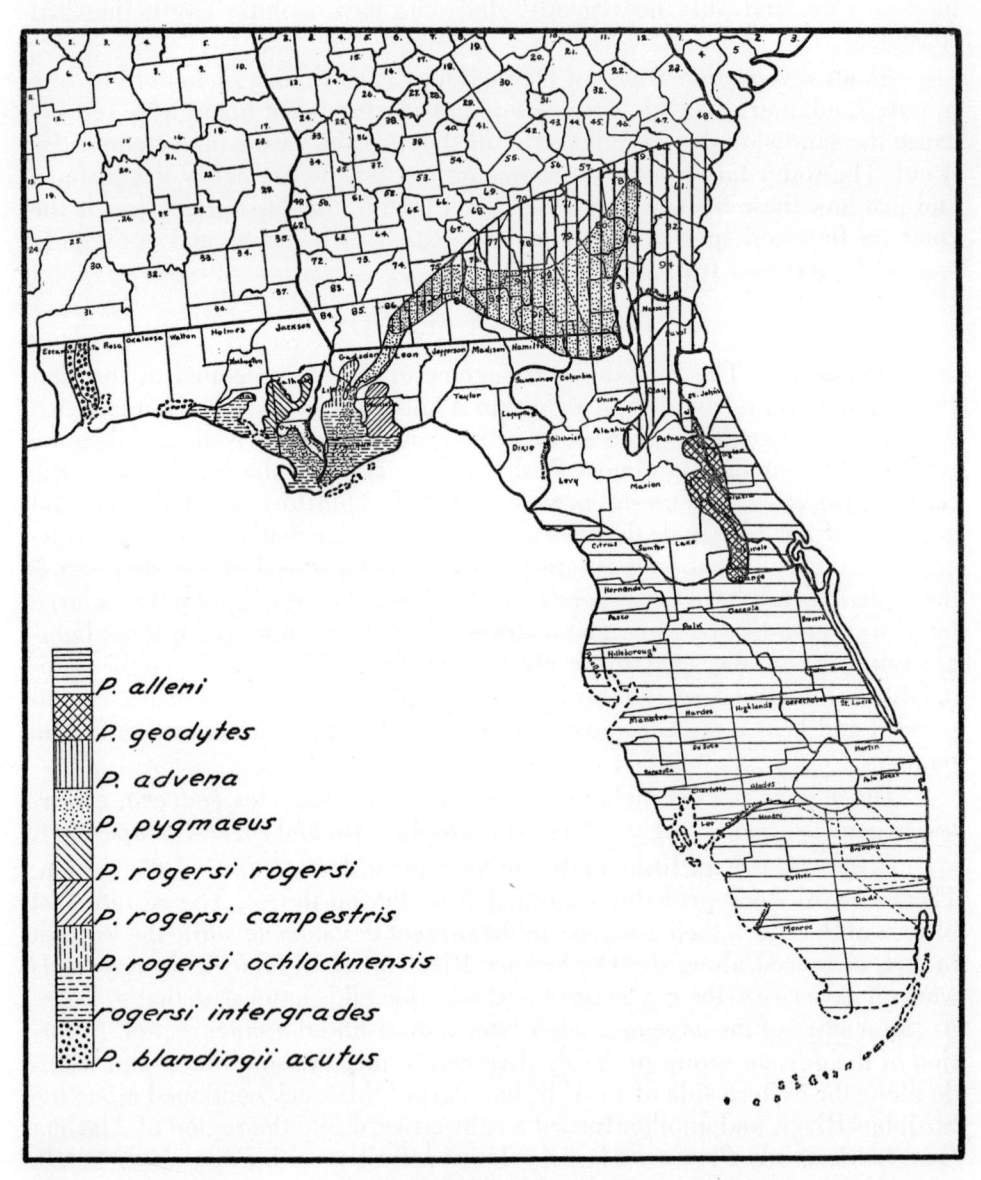

Map 5.—Distribution of the Advena Section, *Procambarus alleni*, and *Procambarus blandingii acutus*.

lacking or is represented by a vestige on the cephalodistal surface; the caudal process is present only as a small swelling—the distinct rim or a corneous process is lacking; the central projection is relatively large and blade-like, compressed laterally, and is directed caudodistad. Hooks are present on the ischiopodites of the third, or the third and fourth pereiopods of the male.

Three species belong to this group, namely, *P. advena*, *P. geodytes*, and *P. pygmaeus*. The range of the group in eastern Florida extends from Nassau County, southward along the St. Johns River and its tributaries to the southwestern part of Seminole County; westward it is found abundantly in the flatwoods in Clay County and is known from one isolated locality in Alachua County. In Georgia the range swings westward into Echols, Lowndes, Cook, Colquitt, and Thomas counties, and thence southward into Liberty and Gulf counties in western Florida. Northward in Georgia it extends into Ben Hill County, and eastward into Appling, Wayne, and Bryan counties.

All of the species of the *advena* group are closely related, but each seems to be either geographically or ecologically isolated from the others. *P. geodytes* is apparently geographically isolated from the other two while *P. advena* and *P. pygmaeus* appear to be ecologically separated.

The barrier, if any, which exists between *advena* and *geodytes* lies in the area along the St. Johns River between Green Cove Springs and the Welaka regions. *P. advena* was collected four miles south of Green Cove Springs, Clay County—[U. S. Highway 17]—and *P. geodytes* was taken 14 miles southeast of Palatka, Putnam County—[State Highway 308]—and at Orange Springs, Marion County.

P. advena is a primary burrowing species while *P. pygmaeus* lives in open water and is probably a secondary burrower; hence, even though the ranges of the two species overlap they occupy different habitats. This more or less complete ecological isolation is moreover probably reinforced by the morphological isolation due to the very small size of *pygmaeus*.

Procambarus advena (LeConte)

Plate IV, Figs. 56-60; Map 5

Astacus advena LeConte 1856, Proc. Acad. Nat. Sci. Philad. 7: 402.
Cambarus carolinus Hagen 1870: 87-88, Pl. I, figs. 51-54, Pl. III, fig. 5.
Cambarus advena Faxon 1884: 113, 140, 141; Faxon 1885a: 8, 12, 47-48, 49, 54-56, 58, 158, 173; Faxon 1885b: 358; Ortmann 1902: 277, 279; Harris 1903: 58, 68, 129, 150; Ortmann 1905a: 98, 100, 101, 104; Ortmann 1905c: 438; Faxon 1914: 412; Hobbs 1938a: 65; Hobbs 1940a: 389, 393; Hobbs (in press-a); Hobbs (in press-d).
Procambarus advena Hobbs (in press-c).

Diagnosis.—Rostrum subovate, acute, without lateral spines; areola narrow; male with hooks on ischiopodites of third pereiopods (hooks in some specimens bituberculate); chela of first form male never bearded but bears a cristiform row of large tubercles along inner margin of palm; postorbital ridges terminating cephalad without lateral spines or tubercles; no lateral spines present on carapace. First pleopod of male reaching coxopodite of second pereiopod and terminating in four parts; mesial process slender and acute, directed caudodistad; cephalic process vestigial; caudal process not distinguishable from bulbiform portion of outer part; central projection by far the most conspicuous, is corneous, blade-like, compressed laterally, and directed caudodistad. Annulus ventralis subovate with the greatest length in the longitudinal axis; cephalolateral margins raised; fossa disappearing beneath caudosinistral wall; sinus, not evident cephalad of fossa, arises from sinistral side of fossa, curves gently to midventral line, and cuts the midcaudal margin of the annulus.

Remarks.—For several years I have attempted to associate the type specimen of this species at the Museum of Comparative Zoology with some of my own specimens, hoping that the type locality might be more specifically known. I have examined the type, made measurements, drawings, and photographs of it, and through the kindness of Dr. Fenner A. Chace of the Museum of Comparative Zoology and Mrs. A. F. Carr, Jr. of the University of Florida, most of my specimens have been compared with it, but still, I am uncertain as to where LeConte's specimen was collected.

While there are a number of local variants in the Florida specimens assigned to *advena* the variation is so slight that the locality from which any given specimen has been taken can be determined only if series of the whole region are available for comparison; however, when these variants are compared with Georgia specimens the differences are more pronounced.

Probably the most decided variation is to be noted in the shape of the rostrum. This is comparatively broader in specimens collected near Baxley, Appling County, Georgia, than in specimens from elsewhere over the range. The nearest approach to the Baxley type of rostrum is shown by two specimens collected near Waycross, Ware County, Georgia. Their rostra are not exactly alike, but neither is so broad, and each is more acute than those of the Baxley forms. Specimens from Clinch, Cook, and Echols counties, Georgia, are acute and not markedly different from the Waycross forms. In Nassau and Duval counties, Florida, the rostral ridges of the specimens are gently convergent up to near the tip where they bend more or less sharply together to form an acute apex (the bend may be either angular or slightly rounded). In specimens from Clay County, Florida, the subrostral ridges extend laterad of the rostral ridges almost to the apex, and the rostrum is almost triangular. A few specimens from Putnam County, Florida, show a radical difference in the general trend of the rostral ridges in that they exhibit a mesial invagination just distad of the base, and the rostrum is definitely not triangular. The Alachua County, Florida, forms have rostra which are very much like the Clay County specimens except that the subrostral ridges do not extend so far laterad as do those of the latter.

The areola is fairly constant over the entire range except in the specimens from Putnam County. In this population it is practically obliterated.

First form males are known from only five localities, and in two of these from only a single specimen. The main differences in the few I have seen seem to be in the development of the cephalic process and the direction in which the central projection is extended. These are possibly individual variations.

A rather careful analysis was made of the annulus ventralis of the *advena* females in my collection, and I find at least four types which are clearly distinct: one type from Appling County, Georgia; another from Alachua County, Florida; another from Putnam County, Florida; and finally, the largest group, including all of the others. Naturally the latter group shows a wide range of minor variations. A single female from Echols County, Georgia, has a very distinct annulus ventralis, but since I have only the one specimen from that locality it is possible that this is again an individual variation.

Although these and a few other characters are definitely variable, the great majority of characters are remarkably uniform throughout the species. This is more noteworthy in that this species does not have a high frequency over its range but tends rather to occur in small local concentrations of presumably inbred populations and appears to have comparatively poor powers of dispersal.

Specimens Examined.—I have examined a total of 81 specimens of *Procambarus advena*, only 27 of which were collected in Florida. This species has been found in the following counties in Georgia and Florida: GEORGIA—Appling, Ben Hill, Bryan, Clinch, Colquitt, Cook, Echols, Lowndes, Pierce, Thomas, Ware, and Wayne. FLORIDA—Alachua, Clay, Duval, Nassau, and Putnam.

SEASONAL DATA

	Jan.	Feb.	Mar.	Apr.	May	June	July	Aug.	Sept.	Oct.	Nov.	Dec.
♂ I				1						3		3
♂ II		2		1	6		3		5	7		3
♀		2		10	1	4	1	4		11	2	6
♀ (eggs)				2			1					
♀ (young)												
♂ (immature)											2	
♀ (immature)										1		

Geographical and Ecological Distribution.—The southern limit of the range of *P. advena* appears to be in Alachua and Putnam counties where specimens have been taken from two dissimilar ecological situations. The eastern limit is marked by the St. Johns River, and the western, except at the extreme south end of the range, by Trail Ridge. The Alachua County records are all from a single locality, a seepage area in a garden and woods along Hogtown Creek about two miles north of Gainesville [U. S. Highway 441].

I have collected over a greater part of Alachua County but have not encountered this species elsewhere. Neither do I have specimens from Bradford

County, but I strongly suspect that numerous isolated colonies are widespread in the eastern part of Alachua County and in Bradford County.

In Georgia the northern limits of the range reach into Ben Hill, Appling, and Bryan counties; here it extends almost a hundred miles farther west than it does in Florida, for I have taken this species 9.5 miles southeast of Meigs in Thomas County. It seems probable that *P. advena* is confined to the coastal terraces and the marshy and wet hillsides of the Tifton Upland along the courses of its small streams.

In Nassau County, Florida, *P. advena* was taken from burrows in a roadside ditch in which there were also burrows of *P. pubischelae* and *P. seminolae*. In Georgia it was also taken with the same two species.

P. advena is a primary burrower, spending most of its life in its burrow. The burrows are beautifully constructed and, though rambling, are very elaborate with numerous galleries. They are made either in some plastic soil or in sand underlain by plastic material, and in the latter case the passage through the upper sand is plastered with mud brought up from the deeper part of the burrow. Usually there are several large chambers, some that are interspersed along the runways and others placed at the terminals of the several passages. It is not uncommon to find a chamber with three or four passages leading away from it. In every burrow there is at least one passage which spirals downward a few inches to two or three feet to a "cellar" below the water table. This passage is usually near one of the chimneys of the ramifying structure. I have often found nuts, large pebbles, grass, and sticks in the passages and chambers.

The chimneys of these burrows are always carefully constructed and in their excellent masonry often approach the chimneys of *C. diogenes*. I have seen them extending five or six inches above the ground level and consisting of neatly formed, round pellets about five-sixteenths of an inch in diameter. Often a single crayfish will construct three or four of these chimneys from its single complex of passageways. Some burrows extend from six to eight feet horizontally with as many as four chimneys, and downward to a depth of three feet to the "cellar." Occasionally I have dug burrows in which a chamber about six inches high and six to ten inches in diameter was located above the cellar passageway.

A single soil class does not seem to determine the distribution of the species, for I have found it in clay, sandy clay (so hard that it felt like sandstone to my hands), sandy muck, loam, and even in seepage areas in which a black soil was predominant. Available subsurface groundwater seems to be a *sine qua non* in determining the distribution; streams and permanent open bodies of water seem to be as barren of this species as do the "Black-jack" ridges. Specimens are sometimes frequent along creeks but are confined to seepage areas well above the normal stream level. The water table in most of the localities I have collected is probably variable, and at times may cover the mouth of the burrows. In some localities where this is obviously the case, the groundwater in dry seasons may descend more than two feet below the surface.

Although *P. advena* occurs chiefly in the flatwoods, it cannot be regarded as a typical flatwoods species; it is apparently just as successful in seepage areas along small streams in a region of high pine and rolling hammock lands.

In the single Alachua County locality this species occurs in comparatively large numbers in the lower corner of a garden bordering Hogtown Creek. At one time this plot was a swampy seepage area. The owner constructed small drainage ditches, cleared away the undergrowth of vegetation, and this one-time swamp has been converted into a very rich garden-spot where the crayfish still occur. There are probably a hundred chimneys in this plot, and fresh ones may be seen almost every morning, especially after a rain. It seems that the crayfish are most active at night. They do not harm the plants except by throwing up chimneys on top of young seedlings. There is another inhabited seepage area about fifty to one hundred yards upstream, and occasional chimneys may be seen at other spots along the creek bed for several hundred yards.

The habitat described above is markedly different from the locality from which this species was taken in Putnam County—a pine flatwoods, 11 miles north of Palatka, in which wire grass and deer's-tongue *(Trilisa)* are common. The soil is a sandy muck underlain by a very plastic clay so that after a heavy rain the whole area is water-logged.

This species appears to be most inquisitive; almost any sort of commotion at the mouth of the burrow seems to attract it to the surface of the water table. Recently I have found that in most instances, particularly in dry weather, when the burrow is disturbed the crayfish either retreats into the deep chamber or was already there. If the passage to this chamber is located, the water thoroughly agitated, and then allowed to remain still, the crayfish usually climbs up the passage to the surface of the water. Here its two antennae may be seen whipping to and fro near the surface, and the crayfish can readily be captured. In other instances it has been necessary to dissect the burrow in order to find its occupant. Then, even though the greatest care is exercised, some of the numerous side passages are lost, and the chances are about even that the crayfish will escape. I have found that a frequently successful and time-saving method of collecting is to open the passage to the deeper chamber of several burrows and thoroughly agitate the water in the mouth of each. If each opened burrow is then cautiously approached, the crayfish is likely to be in the process of repairing or inspecting the damage and may be caught with little difficulty. It is necessary to approach the burrow carefully, however, because the slightest sound will cause the crayfish to retreat into the deeper passages. In the rainy season when the crayfish is frequently in one of the smaller chambers or tunnels nearer the surface rather than in the deepest passage, a careful dissection of the entire burrow is often necessary.

In a flatwoods, in Clay County, I found an area, populated by *P. advena*, in which the burrows possessed no chimneys but opened directly into a ditch that had several inches of water in its bottom. The mouths of the burrows were above the water, but since the water was then quite low I believe that the mouths would be below the surface in rainy seasons. This is the only place where I have encountered such burrows for this species.

Numerous copepods occupy the burrows made by *advena*, and occasionally amphipods of the genus *Crangonyx* are found clinging to the crayfish's swimmerets. Further examination of the crayfish reveals many ostracods (six-legged forms, genus *Entocythere*, authority Ward and Whipple 1918) clinging to the pubescent regions of the ventral surface.

It is likely that the burrows of *advena* form a definite microhabitat, characterized by a small but special fauna in which the other species associated with *advena* have found a stable niche.

Procambarus geodytes, sp. nov.

Plate V, Figs. 61-65; Plate XIX; Map 5

Cambarus advena geodytes Hobbs 1937: 154 (nomen nudum).

Diagnosis.—Rostrum without lateral spines, broad, short, and acute; areola linear or obliterated; male with hooks on ischiopodites of third and fourth pereiopods; palm of chela of first form male not bearded but bears a cristiform row of tubercles; postorbital ridges terminating cephalad without spines or tubercles; no lateral spines on sides of carapace. First pleopod of male, form I, reaching coxopodite of second pereiopod when the abdomen is flexed and terminating in four parts; mesial process corneous, slender and blade-like, and directed caudodistad; cephalic process corneous, very small, and located on the cephalolateral margin near tip of appendage; caudal process not present as a corneous ridge or corneous prominence, but made up of the swollen caudodistal terminal of the appendage; central projection, the most conspicuous of the four terminals, corneous, compressed laterally, acute, and directed caudodistad. Annulus ventralis subcylindrical with cephalic margin deeply cleft and high tuberculate lateral ridges; sinus originates on midventral line about one-fourth the length of the annulus from the cephalic margin, turns gently dextrad, somewhat more sharply sinistrad to cross the midventral line, and finally caudodextrad to the midventral line terminating just before cutting the caudal margin of the annulus.

Holotypic Male, Form I—Body subcylindrical, slightly depressed. Abdomen much narrower than thorax (.96-1.40 cm. in widest parts respectively). Width of carapace greater than depth in region of caudodorsal margin of cervical groove. Greatest width of carapace about midway between cervical groove and caudal margin of carapace.

Areola linear, almost obliterated. Cephalic section of carapace about 1.56 times as long as areola (length of areola 39% of entire length of carapace).

Rostrum deeply excavate, almost reaching penultimate segment of peduncle of antennule; margins converging to tip; no lateral spines present. Upper surface of rostrum with a row of setiferous punctations bordering the high marginal ridges. Subrostral ridges well defined and evident in dorsal view almost to tip.

Postorbital ridge well defined, terminating cephalad in small tubercles. Suborbital angle absent; branchiostegal spine heavy but short. No lateral spines on carapace. Surface of carapace punctate dorsad, tuberculate laterad.

Abdomen shorter than carapace (2.60-2.95 cm.).

Cephalic section of telson with one spine in the sinistral and two in the dextral caudolateral corners.

Epistome subovate with a single cephalomedian acute tubercle, and a caudolateral one on either side at base.

Antennules of usual form. No spine present on ventral side of basal segment.

Antennae extending caudad to first abdominal segment; antennal scale small with mesial and lateral sides subparallel, cephalic margin of blade portion sloping. Spine on outer margin very strong.

Chela distinctly depressed, of moderate length and width. Hand with squamous, ciliated tubercles except on upper proximal and lower surfaces. Inner margin of palm with a cristiform row of seven tubercles. A distinct submedian ridge present on upper surface of both fingers. Fingers slightly gaping. Opposable margin of dactyl with five large, corneous tubercles along proximal three-fifths and a row of minute denticles along distal two-fifths, a few scattered ones between the large tubercles. Lateral margin of dactyl with three squamous tubercles at base and a row of setiferous punctations distad of them. Upper surface of dactyl with a submedian ridge flanked by setiferous punctations. Lower surface with sparsely scattered setiferous punctations. Opposable margin of immovable finger with five prominent tubercles on proximal three-fifths and with a large strongly cornified one extending from lower margin at base of distal fourth. A row of minute denticles broken only by the large tubercles extends along entire margin. Lateral margin of immovable finger with a row of about seven setiferous punctations. Upper surface with a distinct submedian ridge flanked by setiferous punctations. Lower surface with setiferous punctations; setae along lateral portion extremely long.

Carpus of first pereiopod one-half as broad as long; a well defined longitudinal groove above, punctate except on mesial surface which is tuberculate—about six prominent ones.

Merus punctate on mesial and lateral surfaces. Upper surface with an irregular row of small tubercles, two somewhat larger ones near distal end. Lower surface with an outer row of seven tubercles and an inner row of ten, inner row more spike-like.

Hooks on ischiopodites of third and fourth pereiopods; hooks on third bituberculate. Bases of coxopodites of fourth pereiopods with very strong compressed outgrowths directed caudoventrad; those on fifth much smaller but acute and directed cephaloventrad.

First pleopod extending to base of second pereiopod when abdomen is flexed. Tip terminating in four distinct parts. Mesial process corneous and blade-like, and directed caudodistad. Cephalic process small, corneous, acute, and situated on the cephalolateral margin near tip. Caudal process not present as a corneous ridge or prominence but made up of the swollen caudodistal portion of the main shaft. Central projection, the most prominent terminal element, corneous, acute, and directed caudodistad. The whole distal portion of the appendage is bent somewhat caudodistad.

Male, Form II.—Differs from the male, form I, only in the reduction of the secondary sexual parts, and the first pleopod has no indication of a cephalic process; inner margin of palm of chela has eight tubercles; spines on telson are mirrored images of those in holotype; there are also other differences in tubercle counts.

Allotypic Female.—Differs from the first form male chiefly in tubercle count on the chela and other spiny portions.

Annulus ventralis subcylindrical with a single small tubercle on either side at middle; cephalic margin deeply cleft and with high tuberculate lateral ridges; sinus originates on midventral line about one-fourth the length of the annulus from the cephalic margin, turns gently dextrad, somewhat more sharply sinistrad to cross the midventral line and finally caudodextrad to the midventral line just before cutting the caudal margin of the annulus.

Measurements.—Male (form I) Holotype: carapace, height 1.17, width 1.40, length 2.95 cm.; areola, width .01, length 1.15 cm.; rostrum, width .42, length .43 cm.; abdomen, length 2.60 cm.; right chela, length of inner margin of palm .55, width of palm .88, length of outer margin of hand 1.85, length of movable finger 1.33 cm. Female Allotype: carapace, height 1.52, width 1.56, length 3.20 cm.; areola, width .01, length 1.23 cm.; rostrum, width .50, length .50 cm.; abdomen, length 2.95 cm.; right chela, length of inner margin of palm .55, width of palm .96, length of outer margin of hand 1.86, length of movable finger 1.31 cm.

Type Locality.—Orange Springs, in the northeastern part of Marion County, Florida. All of the specimens taken here were dug from complex burrows around the margin of the swimming pool.

Disposition of Types.—The male holotype, the female allotype, and a second form male paratype are deposited in the United States National Museum. Of the remaining paratypes one male (form I) and one female are deposited in the Museum of Comparative Zoology; one male (form I) and one female in the University of Michigan Museum of Zoology. Four males (form I) and nine females are in my personal collection at the University of Florida.

Remarks.—Here again I have too few specimens to determine which, if any, variations are peculiar to local populations. Of the six localities represented in my collection, the specimens from near Welaka seem to represent the most highly inbred population. The rostra of these specimens are broader and resemble those of the Baxley specimens of *P. advena*. The shape of the chelae is also distinct.

The pleopod of the male from Palm Springs seems to be rather distinct; however, since I have only the one specimen it is possible that this is only an individual difference. (The cephalic process is present as a small acute spine at the cephalodistal margin of the gonopod; in the other specimens the process is entirely lacking, or may be represented by a rudimentary bump along the distal cephalolateral margin.)

There is little doubt that *P. geodytes* has its closest affinities with *P. advena*. For some time I considered it as a subspecies of *advena*, but upon further study, I am convinced that my intergradation data is too indefinite to call *geodytes* a race of *advena*. (The form which I have ascribed to *P. advena* collected in Putnam County appears to be somewhat intermediate between the two species; however, since I have so few specimens from this region and no first form male, and the habitat from which they were taken is more similar to that occupied by *P. advena*, those specimens in question will be considered a variant of the latter unless further collecting and study will prove them to be otherwise.)

Specimens Examined.—I have examined a total of 42 specimens of *Procambarus geodytes*, from Marion, Putnam, and Seminole counties. In Marion County it was taken from burrows in the swampy area around Salt Springs in the Ocala National Forest and at Orange Springs; in Putnam County it was found around the edge of a small sulfur spring flowing into the St. Johns River about three miles north of Welaka, and along the shore of Lake Crescent, 3.5 miles north of Crescent City; in Seminole County it was dug from burrows around Palm Springs and at Shepherd Springs just north of Sanlando.

My 42 specimens were collected in April, October, November, and December. First form males were taken during October and November. A single mature male, form II, was taken in December, and one female with young and two immature females in April.

Geographical and Ecological Distribution.—This species is endemic to Florida and appears to be confined to the St. Johns River drainage. Although it is known to occur in only Putnam, Seminole, and Marion counties, I believe that subsequent collecting will reveal its presence in Flagler, Lake, Orange, and Volusia counties.

Procambarus geodytes is not associated with any other crayfish, although both *fallax* and *acherontis* have been taken from springs and spring-runs around which *geodytes* burrows.

Like *P. advena*, *geodytes* is a primary burrower, and I have no evidence that its habits are different from those of *advena*. Although ecologically similar, *geodytes* is almost confined to regions around sulfur or mineral springs where the water table is available only a few inches to a foot below the surface. The burrows of *geodytes* are equally as complex and elaborate as those of *advena*. Along the bank of a small sulfur spring in Putnam County numerous chimneys were crowded together in a small seepage area. I dug into the midst of a group of these and struck a buried log. When this was moved three crayfish were uncovered. Several times I have noticed that members of this species construct runways beneath the entire length of a felled log. One of the largest burrows of this nature I have observed was at Palm Springs; here a board was lying on the ground, and under it was a crayfish tunnel about twelve feet long. At intervals there were vertical passages leading downward from the tunnel. The chimneys constructed by *geodytes* are much like those of *advena*, and they are usually found in regions where the soil is a sandy muck containing an abundance of decaying organic matter.

From burrows at both Orange Springs and Welaka specimens of this species were collected on which amphipods of the genus *Crangonyx* were holding to the abdomens of the crayfish.

Procambarus pygmaeus, sp. nov.

Plate V, Figs. 66-70; Plate XX; Map 5

Diagnosis.—Rostrum without lateral spines, sublanceolate; areola narrow; male with hooks on ischiopodites of third pereiopods only; palm of chela of first form male not bearded but bears a cristiform row of tubercles; postorbital ridges terminating cephalad without spines or tubercles; no lateral spines on sides of carapace. First pleopod of first form male reaching coxopodite of second pereiopod when abdomen is flexed, and terminating in three distinct parts; mesial process slender and blade-like, extends caudodistad slightly beyond the central projection; cephalic process absent; caudal process forming a sharp corneous ridge along caudolateral margin; central projection large, thin, plate-like, and directed distad; distal portion of outer part of appendage hardly inflated, whereas in *advena* it is greatly swollen. Annulus ventralis subovate, with a submedian furrow; sinus originates along midventral line about one-third of the total length from the cephalic margin; curves gently sinistrad, somewhat sharply dextrad to midventral line where it turns caudad to cut the midcaudal margin; fossa not so plainly seen as in *advena*.

Holotypic Male, Form I—Body subovate, compressed laterally. Abdomen narrower than thorax (.68-.81 cm. in widest parts respectively). Width and depth of carapace subequal in region of caudodorsal margin of cervical groove. Greatest width of carapace slightly caudad of caudodorsal margin of cervical groove.

Areola narrow (32 times longer than wide) with a few scattered punctations, only one in narrowest parts; sides parallel for some distance in middle. Cephalic section of carapace about 1.92 times as long as areola (length of areola 34.2% of entire length of carapace).

Rostrum flattened above reaching middle of distal segment of peduncle of antennule; margins converging to tip; no lateral spines present. Upper surface of rostrum with scattered punctations. Marginal ridges high. Subrostral ridges well defined and evident in dorsal view to tip of rostrum.

Postorbital ridge well defined, merging abruptly cephalad into carapace without forming tubercles or spines. Suborbital angle absent; branchiostegal spine moderately developed. No lateral spine on carapace, a few small tubercles instead. Surface of carapace punctate dorsad, granulate laterad.

Abdomen slightly longer than thorax (1.95-1.87 cm.).

Cephalic section of telson with two spines in each caudolateral corner.

Epistome broadly ovate but slightly angular cephalad and cephalolaterad.

Antennules of usual form. No spine present on ventral side of basal segment.

Antennae extending caudad to second abdominal segment; antennal scale small and narrow with cephalic margin of blade portion sloping. Spine on outer margin very strong.

Chela distinctly depressed, of moderate length and breadth. Hand tuberculate except on ventromesial and ventrolateral surfaces. Lower surface of palm with a group of tubercles in two irregular rows running obliquely proximolaterad to mesiodistad. Inner margin of palm with a cristiform row of eight tubercles. A distinct ridge present on upper surface of both fingers. Fingers gaping. Opposable margin of dactyl with six tubercles along basal three-fifths; a single row of minute denticles between these, and continuous to tip. Lateral margins of dactyl with six large tubercles along basal half, distad of which is a row of setiferous punctations. Upper surface of dactyl with a submedian ridge flanked proximad by a group of tubercles, distad by setiferous punctations. Lower surface setose with setiferous punctations. Opposable margin of immovable finger with four tubercles on proximal half and one tubercle extending from lower surface at base of distal third. Crowded denticles interspersed between these. Lateral margin with a row of setiferous punctations. Upper surface with a median ridge flanked by setiferous punctations. Lower opposable surface with a distinct excavation and setiferous punctations.

Carpus of first pereiopod longer than wide (.50-.41 cm.); a shallow longitudinal groove above. Punctate except on mesial surface which is tuberculate. Two tubercles larger than others; a row of three or four tubercles on cephalomesial margin; two tubercles on cephaloventral margin.

Merus punctate on mesial and lateral surfaces. Upper surface with an irregular row of tubercles. Lower surface with an irregular outer row of about 14 tubercles and an inner row of about 12 spike-like tubercles.

Hooks on ischiopodites of third pereiopods only. Bases of coxopodites of fourth pereiopod with a large tubercle directed ventrad; that on fifth very small.

First pleopod extending to base of second pereiopod when abdomen is flexed. Tip terminating in three distinct parts. Mesial process slender and blade-like, directed caudodistad and extends slightly beyond central projection. Cephalic process absent. Caudal process forming a distinct corneous ridge on caudolateral margin at base of central projection. Central projection large and plate-like, directed distad. Distal portion of appendage only slightly inflated.

Male, Form II.—Differs from the male, form I, only in the reduction of the secondary sexual parts and in the first pleopod. Caudal process of first pleopod absent; both the others are somewhat truncate, heavier, and non-corneous.

Allotypic Female.—Differs from the first form male in tubercle count on cheliped, and in proportions of chela; antennae extend caudad to fifth segment of abdomen.

Annulus ventralis subovate, with the greatest length in the longitudinal axis; a longitudinal furrow present along midventral line with high lateral walls along cephalic half; sinus originates on midventral line about one-third of total length from cephalic margin, curves gently sinistrad, then somewhat more sharply caudad to midventral line, where it turns caudad to cut the midcaudal margin of the annulus.

Measurements.—Male (form I) Holotype: carapace, height .78, width .81, length 1.87 cm.; areola, width .02, length .64 cm.; rostrum, width .28, length .34 cm.; abdomen, length 1.95 cm.; right chela, length of inner margin of palm .40, width of palm .52, length of outer margin of hand 1.12, length of movable finger .60 cm. Female Allotype: carapace, height .72, width .80, length 1.80 cm.; areola, width .01, length .61 cm.; rostrum, width .25, length .32 cm.; abdomen, length 1.93 cm.; right chela, length of inner margin of palm .40, width of palm .52, length of outer margin of hand 1.02, length of movable finger .63 cm.

Type Locality.—About 16 miles north of Fargo on Georgia State Highway 89, Clinch County, Georgia. A small swamp stream flowing through a cypress bay in the flatwoods region. *Pontederia sp.*, and *Juncus repens* were abundant in the edge of the stream and in the adjoining roadside ditch. Most of the specimens were taken by pushing a dip net through the dense *Juncus* beds.

Disposition of Types.—The male holotype, the female allotype, and a second form male paratype are deposited in the United States National Museum. Of the remaining paratypes one male (form II) and one female are deposited in the Museum of Comparative Zoology; one male (form II) and one female in the University of Michigan Museum of Zoology. One male (form I), four males (form II), 15 females, and 35 immature specimens are in my personal collection at the University of Florida.

Remarks.—With the few specimens available it is difficult to determine the extent of the local variations in this species. The rostra of the specimens from Clinch County, Georgia, seem to be more lanceolate than those from Wayne County, Georgia, or those from Florida. The basal segments of the antennule of the Clinch County and Florida specimens bear no spine on the ventral surface, while a well developed one is present on that of some of the specimens from Wayne County. These two differences are the most pronounced; however, other and rather minute differences occur, and perhaps with the acquisition of more specimens throughout the range, definite local variants can be recognized.

Procambarus pygmaeus has its closest affinities with *Procambarus advena*. This is particularly interesting since the ranges of these two species overlap considerably. The two are somewhat different, however, in their habits as is pointed out in the discussion of the geographic and ecological distribution of the two species.

Specimens Examined.—I have examined a total of 110 specimens of *Procambarus pygmaeus*, from the following counties in Georgia and Florida. GEORGIA—Clinch and Wayne counties. FLORIDA—Gulf and Liberty counties. The exact localities are: GEORGIA, Clinch County—7.6 miles north of Fargo [State Highway 89]; 15.7 miles north of Fargo [State Highway 89]; 5 miles northeast of Homerville [U. S. Highway 84]. Wayne County—.5 mile north of Jessup [State Highway 38]. FLORIDA, Gulf County—11.7 miles west of Weewahitchka [State Highway 52]; 6.6 miles east of the Gulf-Bay County line [State Highway 52]; 6.1 miles south of Weewahitchka [State Highway 6]; 1.7 miles east of the Gulf-Bay County line [State Highway 52]; 4.1 miles west of Weewahitchka [State Highway 52]. Liberty County—5.4 miles south of Telogia [State Highway 135]; 11.3 miles south of Telogia [State Highway 135].

SEASONAL DATA

	Jan.	Feb.	Mar.	Apr.	May	June	July	Aug.	Sept.	Oct.	Nov.	Dec.
♂ I			1	1						2		
♂ II					5					7	2	
♀			3	1	8			1		17	12	1
♀ (eggs)					1							
♀ (young)												
♂ (immature)				1	5			2		21	1	1
♀ (immature)				1	3					12		1

Geographical and Ecological Distribution.—*Procambarus pygmaeus* is undoubtedly much more widely distributed than is indicated by the few localities from which it has been taken. It is probably scattered throughout the southeastern part of Georgia south of the Altamaha River and in the northern part of Florida in the Okefenokee region. Although this species has not actually been recorded from the northern part of Florida it is very likely that collections made in the northern part of Baker and Columbia counties will disclose its presence there. In the Florida panhandle it is probably common in local areas in Liberty, Franklin, Gulf, and Bay counties.

The range of *P. pygmaeus* as indicated on Map 5 is interesting in that it provides additional support for the conjectured path of migration for at least two other stocks migrating into the Florida panhandle through the Ochlocknee drainage system. It has been postulated (Hobbs, in press-d) that the forerunner of the *rogersi* subspecies had its origin in southeastern Georgia and migrated along the Ochlocknee River into Florida. Further, it is highly probable that the precursor of *P. youngi* utilized this same path. Since *pygmaeus* occurs both in southeastern Georgia and in the western part of Florida, and the apparent connections between these two portions of the range are in the Ochlocknee region, it would seem to add some degree of validity to the above mentioned postulated migration route taken by the *rogersi* and *youngi* stocks.

In Georgia *P. seminolae* and *P. pubischelae* were taken with *P. pygmaeus* in both Clinch and Wayne counties, and in Wayne County *P. advena* was found burrowing near the ditch where *P. pygmaeus* occurred. In Florida *P. kilbyi*, *P. pycnogonopodus*, *P. latipleurum*, and *P. paeninsulanus* were collected with *P. pygmaeus* in Gulf County, and *P. kilbyi* and *P. leonensis* were taken with it in Liberty County.

This species was first taken by Dr. Frank N. Young and myself while collecting at night with headlights, in a swamp stream about seven miles north of Fargo, Clinch County, Georgia. Dr. Young was using a coffee sieve for catching water beetles, and in one scoop dipped up a small crayfish which proved to be an adult female of *P. pygmaeus*. It was about an inch long and marked with brilliant red splotches over a green basal color After this specimen was bagged, we spent some time trying without success to secure a first form male. However, about 16 miles north of Fargo we found the species more abundant and succeeded in capturing about 50 specimens, two of which were first form males.

A glance at a specimen of *pygmaeus* would give the impression that it was a very small, highly colored specimen of *P. advena*. Finding this new species was a surprise, for it had been only a short distance back that I had dug *P. advena* from burrows in a roadside ditch, and to find two so nearly related forms so close together was at least unusual. Even more astounding was the fact that *pygmaeus* was not dug from burrows but was taken from flowing water. Judging by the method used in collecting them, they must have been out crawling under the vegetation and over the bottom of the ditch and stream. Since most of my collecting was done at night I was unable to ascertain whether these specimens had burrows in the bottom of the ditch and stream or whether they were true open water forms.

A number of specimens of *pygmaeus* has been added to my collection since that time, but all of them have come from similar situations. It has been noted that every locality from which this species has been taken is in swampy terrain, and whether in quiet or flowing water *Juncus repens* is abundant. In most of these localities this brilliantly red and green colored plant forms thickly matted beds over large areas of the pool or stream bottoms.

For some time I was unable to find this species burrowing, and this seemed unusual in view of the fact that its relatives are all primary burrowers. Specimens have now been taken from burrows in nearly all of the localities cited. *P. pygmaeus* apparently is a secondary or tertiary burrower. These burrows are fully as complex as those of the other members of the *advena* section, having a number of side passages and sometimes several openings over which are moderately well constructed chimneys. Most of the burrows I have seen were in soft muck very close to the edge of the water or in recently dried up ditches.

THE ROGERSI GROUP

Diagnosis.—Rostrum broad, short, and devoid of lateral spines; areola very narrow or obliterated; male with hooks on ischiopodites of third pereiopods only; chela compressed and bearing a cristiform row of tubercles along inner margin of palm; postorbital ridges terminating cephalad without lateral spines; no lateral spines present on sides of carapace. First pleopod of first form male terminating in three or four parts; mesial process well developed, spiniform, or slightly compressed; cephalic process present or absent, if present consists of a reduced spine on cephalodistal surface; caudal process large and thumb-like, bent mesiad at a 15-90 degree angle to the main shaft; central projection large and plate-like, extending across cephalodistal surface or obliquely distad in a cephalomesial to caudolateral direction.

A summary of the relationships and distribution of this group given by Hobbs (in press-d) is quoted in part: "The *rogersi* group consists of three distinct though intergrading races, the ranges of which cover a considerable area in the eastern and central parts of the panhandle of Florida. The extreme eastern and western limits of the range of the complex are almost one hundred miles apart, while the most northern and southern limits are separated by a distance of about fifty miles. Within these limits are portions of Bay, Gadsden, Gulf, Calhoun, Franklin, Leon, Liberty, and Wakulla counties.

"All of the members of the complex are primary burrowers with presumably poor powers of dispersal. Their ranges extend through a monotonous flatwoods, broken only here and there by small, scattered areas unsuitable for habitation by them. If *rogersi* had a higher vagility such a range would seem to favor maintenance of a homogeneous population over the entire area; but instead the region is inhabited by small, local populations, and when specimens from several of these are compared, it becomes evident that the *rogersi* complex, especially in the zone of intergradation, is very heterogeneous. In a series of specimens from the region of intergradation, transitional forms between the three subspecies may be beautifully traced.

"The ranges of the three races . . . [are briefly indicated below]. The zone of intergradation consists of a large area in Franklin, Gulf, Bay and the southern parts of Liberty and Wakulla counties. Toward the eastern side of this area, in eastern Franklin and western Wakulla counties, specimens show a definite approach to *rogersi campestris*, which occurs in Leon and Wakulla counties; west of the Apalachicola River, in northern Gulf County, material is closer to *rogersi rogersi*; while in the southern part of Liberty County the intermediates in most characters more closely resemble *rogersi ochlocknensis*. In general, the nearer one approaches the ranges of the three well defined subspecies, the more nearly do the intermediates resemble typical material of these subspecies. The actual situation in respect to the intergrades is more complex than is here indicated . . . "

Procambarus rogersi rogersi (Hobbs)

Plate V, Figs. 71-75; Map 5

Cambarus rogersi Hobbs, Journ. Wash. Acad. Sci. 28 (2): 61-65, figs. 1-11.
Cambarus rogersi Hobbs 1940a: 410; Hobbs (in press-a).
Procambarus rogersi Hobbs (in press-c).
Procambarus rogersi rogersi Hobbs (in press-d).

Diagnosis.—Rostrum without lateral spines; areola very narrow, almost obliterated; male with hooks on ischiopodites of third pereiopods only; chela compressed and bearing a cristiform row of tubercles along inner margin of palm; postorbital ridges terminating cephalad without spines or tubercles; no lateral spines present on carapace. First pleopod terminating in three or four parts; mesial process well developed, spiniform; cephalic process usually absent, if present consisting of a small spine on cephalodistal surface; caudal process non-corneous, thumb-like, and bent caudomesiad at a 90 degree angle to the main shaft; central projection corneous, plate-like, and directed across the cephalic tip of the appendage. Annulus ventralis subovate, broader than long with cephalic margin entire; lateral walls without tubercles.

Remarks.—This subspecies is confined to a relatively narrow strip of flatwoods in Calhoun County, where each of the local populations constructs a group of complex burrows.

Specimens Examined.—I have examined a total of 36 specimens of this subspecies, collected during April, May, June, and November. First form males were collected in April and May, and one female carrying eggs was taken in April.

Procambarus rogersi ochlocknensis Hobbs

Plate V, Figs. 76-80; Map 5

Procambarus rogersi ochlocknensis Hobbs (in press-d).

Diagnosis.—Rostrum without lateral spines; areola obliterated; male with hooks on ischiopodites of third pereiopods only; chela compressed and bearing a cristiform row of tubercles along inner margin of palm; postorbital ridges without spines or tubercles; no lateral spines present on carapace. First pleopod terminating in four parts; mesial process well developed, spiniform; cephalic process forming a spine on cephalodistal surface; caudal process heavy, non-corneous, somewhat inflated and directed mesiodistad; central projection corneous, thin, plate-like, and directed obliquely caudolaterad from the cephalodistal surface. Annulus ventralis subovate, distinctly longer than broad, no cephalic wall (cephalic margin deeply cleft); lateral walls bearing small tubercles.

Remarks.—*P. rogersi ochlocknensis* is found in seepage areas along tributaries of the Ochlocknee River in Gadsden County, and from similar situations and flatwoods in the northern parts of Liberty County. Like the other two subspecies it seems to be largely restricted to local areas where ground water is available within two or three feet below the surface at all times of the year. All of the burrows of this subspecies are extremely complex, consisting of intricate tunnels and at least one deep passage which spirals downward.

Specimens Examined.—I have examined a total of 49 specimens of *P. rogersi ochlocknensis*, which were collected in March, April, May, August, and December. First form males were taken in March, April, May, and December, and two females with eggs were found in March.

Procambarus rogersi campestris Hobbs
Plate VI, Figs. 81-85; Map 5

Procambarus rogersi campestris Hobbs (in press-d).

Diagnosis.—Rostrum without lateral spines; areola obliterated; males with hooks on ischiopodites of third pereiopods only; chela compressed and bearing a cristiform row of tubercles along inner margin of palm; postorbital ridges without spines or tubercles; no lateral spines present on carapace. First pleopod terminating in four parts; mesial process well developed, spiniform; cephalic process forming a spine on cephalodistal surface; caudal process thumb-like and extends mesiodistad at a 45 degree angle to the main shaft; central projection forming a large corneous fan across cephalic side of tip and bent laterodistad at about a 45 degree angle to the main shaft. Annulus ventralis subcylindrical with cephalic wall deeply cleft; lateral walls tuberculate.

Remarks.—Typical specimens of this subspecies are found in the southwestern part of Leon County and in the northwestern part of Wakulla County. All of my specimens have been taken from complex burrows in the low flatwoods areas in this region.

Specimens Examined.—I have examined a total of 37 specimens of *P. rogersi campestris*, which were collected in May, June, August, and November, and first form males were taken in June and November.

The Rogersi Intergrades

Intergrades between the three subspecies of *Procambarus rogersi* occupy an area larger than the combined ranges of the three defined races and extend considerably to the west of the range of *P. rogersi rogersi*. The area of intergradation extends from the southern part of Liberty County westward through Franklin, Gulf, and Bay counties west of St. Andrews Bay.

As has been pointed out by Hobbs (in press-d) the main characters which furnish the intergradation data are the first pleopod of the male and the annulus ventralis of the female.

Specimens Examined.—I have examined a total of 192 *rogersi* intergrades, which were collected during the months of April, May, June, September, October, November, and December. First form males were found in May and December, and six females with eggs in May and one female with young in June.

THE ACHERONTIS SECTION

A separate *acherontis* section is necessitated to receive the very disjunct *Procambarus acherontis*.

Procambarus acherontis (Lönnberg)

Plate VI, Figs. 86-90; Map 8

Cambarus acherontis Lönnberg 1894b, Bihand till K. Svenska Vet.—Akad. Handlingar, Vol. 20, Pt. 4, No. 1: 6-12, Figs. 1-5b.

Cambarus acherontis Lönnberg 1894a: 125-127; Hobbs 1940a: 387-394, 401, fig. 15.

Procambarus acherontis Hobbs (in press-c).

Diagnosis.—Rostrum with small lateral spines; areola narrow (about 20 times as long as broad); male with bituberculate hooks on third and fourth pereiopods; palm of chela not bearded within; postorbital ridges terminating cephalad with spines; lateral spines present on sides of carapace; eyes reduced and without pigment, and the body albinistic. Cephalodistal surface of first pleopod of male not terminating in a ridge or a knob-like prominence but almost flush with the central projection; cephalic process absent; mesial process lies along the caudomesial margin and terminates one and one-half times its own length proximad of the tip; central projection strongly developed, subtriangular, corneous, flattened laterally, and extends beyond the rest of the appendage distally. Annulus ventralis with cephalic portion hidden beneath two projections from the sternum just cephalad of it; subovate in general outline with a lyre-shaped prominence above; sinus originates on midventral line near base of the above mentioned prominence, curves gently dextrad until it approaches the caudal margin of the annulus where it turns sharply sinistrad and finally caudad where it cuts the caudal margin just sinistrad of the midventral line. (This structure shows considerable variation, especially in the bending of the sinus, the general outline, and the presence or absence of tubercles on the cephalolateral surfaces.)

Remarks.—Among the specimens I have at hand there is considerable variation. All spiny parts show considerable diversity; the rostrum has two or three extra spines in addition to the usual two, and the telson also bears a variable number of spines. The annulus ventralis is the most non-stable structure I observed; not only is it different from specimen to specimen in shape, but the sinus takes a number of different forms. With all of these variations, however, none are so great as to confuse *acherontis* with any other known species.

This species has no close affinites with any other crayfish, but I believe it to be more nearly related to the members of the *advena* group than to any other. In some respects it appears intermediate between the *spiculifer* and *advena* groups. The lateral spines on the rostrum and bituberculate hooks on the fourth pereiopods are like those of *P. versutus*, but the other characters are definitely more like those of *advena*.

As a result of the fact that the type specimens of *P. acherontis* were no longer extant at the time that Faxon examined troglodyte forms from Gum Cave, Citrus County, Florida, he mistook them to be specimens of *P. acherontis*. Later authors accepted Faxon's identifications, and this precluded the possibility of Ortmann's arriving at a correct evaluation of the true relationships of *acherontis* as well as leading to a series of misidentifications of Florida's cavernicolous crayfishes. In at least two instances these misidentifications have been recorded in literature. Lönnberg's *Cambarus acherontis* was lost from the time it was described in 1894 until November 1938 (Hobbs 1940a: 338).

Specimens Examined.—I have examined a total of 44 specimens, from Seminole County, Florida. All of these specimens were taken from Palm Springs, about 12 miles north of Orlando on November 11, 1938. [Among them are 3 ♂ ♂ (form I), 13 ♂ ♂ (form II), 12 ♀ ♀, 9 ♂ ♂ (immature), and 7 ♀ ♀ (immature).]

Geographical and Ecological Distribution.—As has been pointed out elsewhere, this species is probably a relict of a once more widely spread stock which migrated into Florida soon after the first closing of the Suwannee Straits. Then with submergence of the land bridging the Straits and the possible inundation of the whole island, a portion of this stock retreated into subterranean freshwater. Whether or not the whole peninsula was submerged after the reopening of the Suwannee Straits, it does seem highly probable in light of its aberrant structures and distribution that *acherontis* was derived from the earliest surviving stock to arrive in the peninsular region of the state.

No published observations other than those of Lönnberg have been made on the habits of *P. acherontis* except those I made at Palm Springs in 1938. "We found more than a score of white crayfish lying in the algae over the bottom of a pool formed by the spring. This pool is about 60 by 20 feet and over the most part approximately six feet deep. The walls and bottom were covered with a thick algal growth, and deposited on it was a sediment characteristic of sulfur springs. The water had a pH of 7.6. Mr. Marchand caught most of the forty-four specimens which were secured by diving to the bottom and capturing them with his hands. They were extremely sluggish, many lying in the algae on their backs with their feet turned up toward the surface as though dead. Even after they were bagged, there was appreciably little sign of life" (Hobbs 1940a: 338).

Lönnberg's account points out the fact that this species does not confine itself to the mouths of springs or sinks but occurs in at least one underground stream, evidenced by his specimens found in a well dug near Lake Brantley. "At first they were fairly numerous, but later on, when I had heard about it and tried to obtain some specimens, I could only procure two males" (Lönnberg 1894b: 4).

THE BLANDINGII SECTION

Diagnosis.—Cephalodistal margin of first pleopod of male never bears a ridge or knob-like prominence unless it is a part of one of the terminal processes; distad the appendage may be directed straight or caudad; mesial process generally bent (either caudodistad or caudolaterad); a crescentric terminal protuberance never present; cephalic process when present arises from cephalic or cephalolateral margin, never from mesial surface; hooks present on ischiopodites of third and fourth pereiopods.

Members of this section are found throughout the coastal regions from Texas to New Jersey and throughout the middle west. They are found in almost any type of aquatic situation and frequently construct burrows.

BLANDINGII SECTION

Blandingii Group
    Blandingii Subgroup
    Clarkii Subgroup
    Evermanni Subgroup
    Fallax Subgroup
Spiculifer Group
Pictus Group
    Pictus Subgroup
    Lucifugus Subgroup
    Seminolae Subgroup

THE BLANDINGII GROUP

The blandingii group consists of four subgroups (the *blandingii, clarkii, fallax,* and *evermanni*) which constitute a very heterogeneous assemblage. While the former three subgroups have rather distant affinities with each other, apparently having been developed along different lines, they are tied together by definite affinities with *evermanni*. Because of this a definition of the *blandingii* group would involve a combination of the descriptions of the four subgroups. The only characters common to all of the members of the *blandingii* group are: hooks on the ischiopodites of the third and fourth pereiopods; a relatively narrow or obliterated, long areola and a complex of general similarities that are easier to see than to precisely define. Among these is the structure of the rostrum which may or may not bear lateral spines; if lateral spines are absent, then the margins are always interrupted. The areolae of the *spiculifer* and *pictus* groups are broad and short except in the cavernicolous species of the latter.

THE BLANDINGII SUBGROUP

Diagnosis.—The first pleopod terminates in four well developed parts; no hump is present on the anterior margin of the appendage; mesial process subspiculiform or narrowly blade-like; cephalic process never spiculiform, is hood-like or blade-like; central projection conspicuously large and corneous; corneous caudal process strongly developed, lies caudad of the central projection, and does not obscure the centrocaudal process of the latter. Margins of the rostrum are broken and often bear spines; areola long and relatively narrow.

The *blandingii* subgroup comprises six species and subspecies, only two of which occur in Florida: *blandingii blandingii, blandingii acutus, blandingii cuevachicae, hayi, lecontei,* and *bivittatus.*

The range of the *blandingii* subgroup is the same as that of the *blandingii* section.

Procambarus blandingii acutus (Girard)

Plate VI, Figs. 91-95; Map 5

Cambarus acutus Girard 1852, Proc. Acad. Nat. Sci. Philad. 6: 91.
Cambarus acutissimus Girard 1852: 91.
Cambarus stygius Bundy 1876: 3.
Cambarus blandingii var. *acuta* Faxon 1884: 136.
Cambarus blandingii acutus Faxon 1890: 619.
Procambarus blandingii acutus Hobbs (in press-c); Hobbs and Marchand (in press).

A complete bibliographic citation does not seem desirable in the case of this subspecies for at least two reasons. One is that I am not certain that the Florida specimens I have referred to *P. blandingii acutus* should be placed here. Although this crayfish has been treated numerous times in literature, no thorough taxonomic consideration has been given it. Faxon (1885: 20-23) presented what was known at that time, but with the accession of numerous specimens in the several museums and private collections, his picture is only fragmentary at best. This *blandingii* complex with three described subspecies *(blandingii, acutus,* and *cuevachicae)* is not at all understood; further specimens which answer the descriptions of the first two are found in overlapping regions and could be ascribed to either subspecies.

The other reason for not attempting a complete bibliographic citation is because of the unreliability of references in literature. It is not at all certain what species or variety the author was referring to when he pointed out the habits of the animal in question. It goes without stating that this species needs considerable work, not only a comparison of specimens in collections but the collecting of large series from all parts of its range, which extends from the eastern states throughout the Mississippi Valley into Mexico.

Diagnosis of Florida specimens ascribed to this subspecies.—Rostrum broad at base, tapering; small lateral teeth present near tip; tip reaching base of distal segment of peduncle of antennule; areola narrow with one or two punctations in narrowest part; male with hooks on ischiopodites of third and fourth pereiopods, hooks long and slender; chela of first form male never bearded along inner margin of palm, instead there is a row of eight or nine tubercles; chela long and slender; opposable margin of movable finger convex; postorbital ridges terminating cephalad in acute spines; one lateral spine present on each side of carapace. First pleopod of male reaching coxopodite of third pereiopod; tip terminating in four distinct parts; mesial process long and spiniform, and directed laterodistad; cephalic process blade-like, corneous, compressed laterally, and directed caudodistad; caudal process and central projection acute, blade-like, corneous, compressed laterally, and directed caudodistad; a terminal tuft of setae borne on a laterally situated knob-like prominence. Annulus ventralis broadly subovate with the greatest length in transverse axis; sinus originates slightly caudad of cephalic margin and dextrad of midventral line, curves gently caudosinistrad crossing the midventral line; a short distance cephalad of the caudal margin it turns caudodextrad to end abruptly just before it cuts the midcaudal margin of the annulus.

Remarks.—It was not until the summer of 1941 that this species was found in Florida. Mr. Ray Boles, who was collecting specimens for me in the northern part of Santa Rosa County, brought me one second form male and four females. Two trips have been made to this region since that time, and I now have a small series of specimens from that region. I had collected it from Baldwin County, Alabama, but until recently had been unable to find this species in Escambia County, Florida, in spite of several searches.

P. blandingii has its closest affinities with *P. hayi, P. lecontei,* and *P. bivittatus.*

Specimens Examined.—I have a total of 145 specimens from Florida, collected in Santa Rosa and Escambia counties. The localities are as follows: about 12 miles west of Jay along McCaskill's Mill Creek, Santa Rosa County (two localities in this region about two miles apart); and a small sand bottomed creek at Century, Escambia County. Thirty-seven first form males were taken during May.

Geographical and Ecological Distribution.—*Procambarus blandingii acutus* has been reported from the following states: Alabama, Arkansas, Iowa, Indiana, Kansas, Louisiana, Michigan, Mississippi, Missouri, North Carolina (?), Ohio, Oklahoma, South Carolina (?), Tennessee, Vera Cruz, Mexico (?), and Wisconsin. As has been pointed out above, this species is poorly known, and certainly the localities cited for South Carolina, North Carolina, and Vera Cruz, Mexico need confirmation.

This subspecies has been found in Florida only along the Escambia River drainage system, although it is probable that it will also be found in the Perdido River system. It is worthy of note that three other species not found elsewhere in the state have been taken in this same drainage *(P. shermani, P. bivittatus,* and *C. species incertis).* In addition *C. diogenes, P. versutus, P. evermanni, P. okaloosae,* and *C. schmitti* have been found here. All of these species except *C. schmitti* and *P. evermanni* were taken in at least one of the three localities with *blandingii acutus.*

Considerable variations have been noticed in the habits of this subspecies in different parts of its range, and Hobbs and Marchand (in press) have given a summary of the more diverse reports on its ecological preferences. Little can be said concerning the ecology of this species in Florida, for it has been collected from only three localities, and these include two somewhat contrasting habitats. In the Escambia River bottoms along McCaskill's Mill Creek it was taken from shallow sloughs and flood plain pools where the water was turbid. At Century this crayfish was taken along with *P. okaloosae* from a small, clear, sand bottomed creek which has the appearance of being somewhat temporary. In both instances I was collecting at night with the aid of a headlight, and the crayfishes were taken from open water. Aquatic vegetation was extremely sparse in both situations. One specimen of *blandingii acutus* was dug from a shallow, simple burrow in a drying slough along McCaskill's Mill Creek.

Procambarus bivittatus, sp. nov.

Plate VI, Figs. 96-100; Plate XXI; Map 9

Diagnosis.—Rostrum long, with lateral spines; areola very narrow, with one punctation in narrowest part; male with hooks on ischiopodites of third and fourth pereiopods; palm of chela of first form male never bearded within but with a row of seven or eight prominent tubercles; postorbital ridges terminating cephalad in strong spines; a single, large lateral spine present on each side of carapace. First pleopod of male, form I, without a shoulder on cephalic surface, reaching coxopodite of third pereiopod and terminating in four distinct parts; mesial process blade-like with an acute tip and directed caudodistad; cephalic process lamelliform, compressed laterally, caudodistal tip forming an acute angle; caudal process corneous, lamelliform, subtriangular, acute, and directed cephalodistad; central projection corneous, slender, triangular, and directed caudodistad. Annulus ventralis egg-shaped with the narrow end cephalad; sinus originates on midcephalic border, extends caudad along midventral line, turns slightly sinistrad, and makes a hairpin turn back to midventral line, proceeding along the latter almost to caudal margin of annulus.

Holotypic Male, Form I.—Body subovate, compressed laterally. Abdomen narrower than thorax (1.77-2.05 cm. in widest parts respectively). Width of carapace less than depth in region of caudodorsal margin of cervical groove. Thoracic portion of carapace with subparallel margins.

Areola very narrow with only one punctation in narrowest part; cephalic section of carapace a little greater than twice as long as areola (length of areola 31.8% of entire length of carapace).

Rostrum deeply excavate, reaching distal end of peduncle of antennule; margins parallel for a short distance at base but converging to base of acumen, the latter set off by prominent lateral spines. Upper surface of rostrum with a row of setiferous punctations along base of marginal ridges. Marginal ridges prominent. Subrostral ridges moderately well defined but not evident in dorsal view.

Postorbital ridge strongly developed and terminating cephalad in acute spines. Suborbital angle obtuse but prominent; branchiostegal spine strong. A single large lateral spine on each side of carapace. Surface of carapace punctate dorsad, tuberculate laterad.

Abdomen slightly shorter than carapace (4.80-5.03 cm.).

Cephalic section of telson with two spines in each caudolateral corner.

Epistome broadly subovate with a cephalomedian projection.

Antennules of usual form. A strong spine present on ventral side of basal segment.

Antennae extend caudad beyond caudal margin of telson; antennal scale of moderate width, broadest slightly proximad of middle; cephalomesial angle rounded; spine on outer margin well developed.

Chela long and slender, subovate, depressed. Hand with a number of rows of tubercles on all surfaces. Inner margin of palm with a row of seven or eight tubercles. Neither finger with a well defined submedian ridge. Fingers not gaping. Opposable margin of dactyl with nine rounded tubercles on proximal half, between and distad of which are crowded minute denticles. Lateral surface of dactyl with three proximal tubercles distad of which is a row of setiferous punctations. Upper and lower surfaces with a few tubercles on proximal portions, otherwise with setiferous punctations. Opposable margin of immovable finger with 12 rounded tubercles on proximal half and a large one extending from lower margin at base of distal half. Between and distad of these are crowded minute denticles. Lateral, upper, and lower surfaces of immovable finger with setiferous punctations.

Carpus of first pereiopod about 1.5 times as long as broad; a shallow and poorly defined longitudinal groove above, punctate except on mesial and dorsomesial surfaces. Two prominent tubercles on mesial surface, one on dorsomesial distal margin, one on ventromesial distal margin, and one on ventrodistal margin.

Merus punctate on lateral and proximomesial surfaces, tuberculate otherwise. An irregular row of tubercles on upper surface with two larger and more spiniform ones on upper distal portion. Lower surface with an irregular row of about 11 tubercles and an inner row of about 14.

Hooks on ischiopodites of third and four pereiopods, those on fourth bituberculate. Bases of coxopodites of fourth and fifth pereiopods with prominent outgrowths. Those on fourth compressed laterally and those on fifth, cephalocaudad.

First pleopod extending to base of third pereiopod when abdomen is flexed. Tip terminating in four parts. Mesial process blade-like with an acute tip and directed caudodistad. Cephalic process lamelliform, compressed laterally, caudodistal tip forming an acute angle, cephalodistal portion rounded. Caudal process corneous, lamelliform, subtriangular, acute, and directed cephalodistad. Central projection corneous, slender, triangular, and directed caudodistad.

Male, Form II.—Differs from the male, form I, in having a slenderer and more weakly tuberculate chela; hooks on ischiopodites of fourth pereiopods show only the faintest tendency toward being bituberculate; outer portion of first pleopod truncate distad with only one or two small tubercles; lateral margins of epistome tuberculate and forming an angle on midcephalic margin.

Allotypic Female.—Differs from the male, form I, in having a comparatively shorter chela; epistome similar to that of the second form male.

Annulus ventralis egg-shaped with the narrow end cephalad; sinus originates on midcephalic border, extends caudad along midventral line, turns sharply sinistrad and makes a hairpin turn back to midventral line, proceeding along the latter almost to caudal margin of annulus.

Measurements.—Male (form I) Holotype: carapace, height 2.17, width 2.05, length 5.03 cm.; areola, width .05, length 1.60 cm.; rostrum, width .69, length 1.64 cm.; abdomen, length 4.80 cm.; right chela, length of inner margin of palm 1.78, width of palm 1.10, length of outer margin of hand 4.72, length of movable finger 1.65 cm. Female Allotype: carapace, height 2.54, width 2.12, length 5.54 cm.; areola, width .05, length 1.77 cm.; rostrum, width .76, length 1.69 cm.; abdomen, length 5.44 cm.; right chela, length of inner margin of palm 1.27, width of palm 1.03, length of outer margin of hand 3.75, length of movable finger 2.23 cm.

Type Locality.—Sloughs along the Escambia River on State Highway 62, Escambia County, Florida. Here the flood plain is at least a mile wide and is inundated for long periods every spring. When the water recedes, a large number of sloughs and isolated pools are left, and some of the shallower ones which are exposed to bright sunlight support dense growths of *Juncus repens*. Here *P. bivittatus* and *P. blandingii acutus* are abundant.

Disposition of Types.—The holotypic male, the allotypic female, and a second form male paratype are deposited in the United States National Museum. Of the remaining paratypes one male (form I), one male (form II), and one female are deposited in the Museum of Comparative Zoology; one male (form I), one male (form II), and a female in the University of Michigan Museum of Zoology. Two males (form I), 21 males (form II), 51 females, 18 immature males, and 32 immature females are in my personal collection at the University of Florida.

Remarks.—*Procambarus bivittatus* is known only from the Escambia River drainage system, and my specimens show few variations; the slight differences in the number of tubercles, the shape of the epistome and annulus ventralis are not so great as to confuse *bivittatus* with other species.

A distinct longitudinal dark stripe along the dorsolateral surface of the thoracic region of the carapace constitutes an excellent diagnostic character, for the only other species in which such a stripe is so pronounced is *P. lecontei*, of the Mobile, Alabama, region. Although *bivittatus* has its closest affinities with *P. lecontei*, it is closely allied to the other members of the *blandingii* group.

Specimens Examined.—I have examined a total of 133 specimens, from the northern part of Santa Rosa and Escambia counties, Florida. All of my specimens were taken during the months of April, May, and October, and first form males were taken in May.

Geographical and Ecological Distribution.—This species is known only from the Escambia River drainage in the northern parts of Escambia and Santa Rosa counties. Most likely it will be found in the Conecuh and the Escambia rivers in southern Alabama. Specimens have been taken from both large and small, clear, sand bottomed creeks and from muddy sloughs. A few specimens were dug from simple burrows.

THE CLARKII SUBGROUP

Diagnosis.—A *clarkii* group was recognized by Ortmann (1905a: 100) and defined as follows: "Outer part of sexual organs with two compressed tubercles, inner part straight, directed forwards. Anterior margin with a distinct shoulder. Rostrum with marginal teeth, acumen rather short. Areola very narrow, often obliterated in the middle, about half as long as anterior section

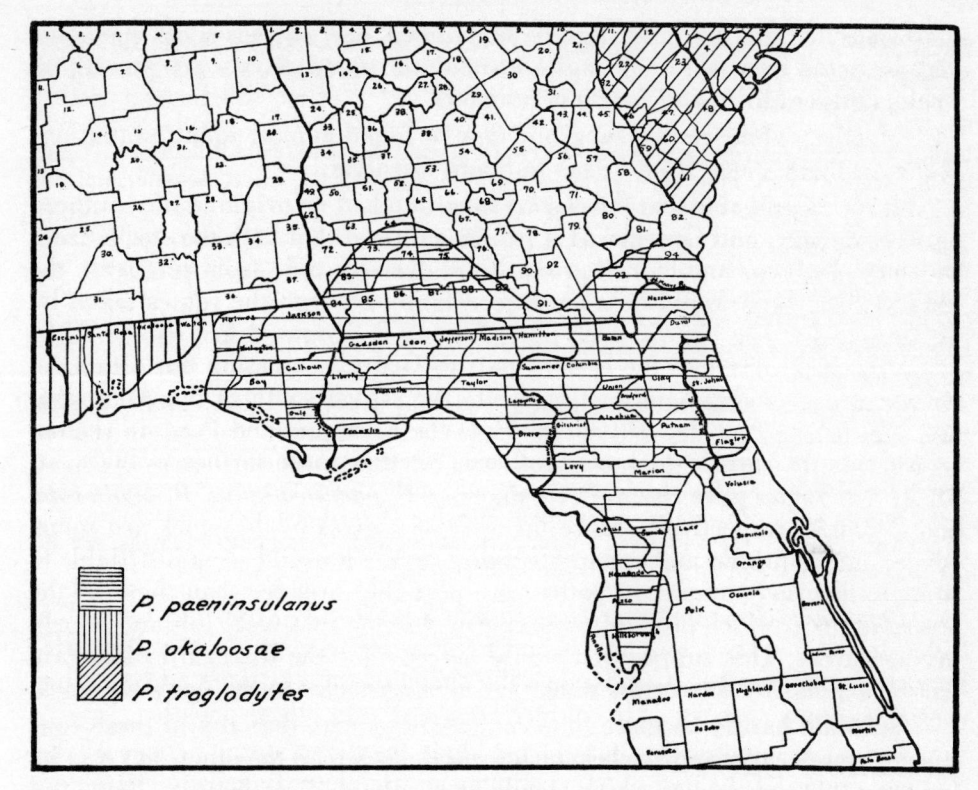

Map 6.—Distribution of the Clarkii Subgroup.

of carapace." In this group he included two species, *clarkii* and *troglodytes*. Two other species, *paeninsulanus* and *okaloosae,* may be added to the subgroup.

The following diagnostic details may be added to the above: first pleopod of male—mesial process spiniform and may be directed distad or caudodistad; cephalic process may be almost blade-like *(paeninsulanus)*, rather long and compressed laterally and terminating caudodistad in a more or less acute angle *(clarkii* and *okaloosae)*, or slightly inflated (sometimes compressed) and the caudodistal margin rounded *(troglodytes)*; caudal process always strongly developed, consisting of a corneous shield-like structure which almost obscures the centrocaudal process of the central projection, extending around it both laterad and caudad (in *paeninsulanus* it is not so shield-like but extends as a flattened blade and lies laterad in apposition to the central projection; caudad it is divided, and a caudomesial part extends caudomesiad of the central projection); central projection is a corneous outgrowth from center of tip and is always compressed laterally, may be partially hooded by cephalic process and is always partially hidden (viewed laterally) by caudal process; decided shoulder present on cephalic margin of appendage. Areola narrow and obliterated; lateral spines sometimes present on rostrum;

Merus punctate on lateral and proximomesial surfaces, tuberculate otherwise. An irregular row of tubercles on upper surface with two or three larger and more spiniform ones on upper distal portion. Lower surface with an irregular row of about 13 tubercles and an inner row of about 15.

Simple hooks on ischiopodites of third and fourth pereiopods. Bases of coxopodites of fourth and fifth pereiopods with strong outgrowths; those on fourth rounded and heavy; those on fifth compressed and much smaller.

First pleopod extending to base of third pereiopod when abdomen is flexed. Tip terminating in four parts. Mesial process spiniform, tip corneous, and directed caudodistad. Cephalic process with a corneous edge, compressed laterally, and the largest of the terminal elements. Caudal process corneous and forming an undulate plate across the caudal and caudolateral portion of the outer part. Central projection small, corneous, triangular, compressed cephalocaudad, and hardly distinguishable from the cephalolateral part of the caudal process. Cephalic margin of left pleopod with a distinct shoulder; right pleopod with a shoulder folded caudomesiad and lying along the cephalomesial surface.

Male, Form II.—Differs from the male of the first form in having a slenderer and slightly longer rostrum; the chelae slenderer with weaker tubercles; cephalic section of telson with four spines in each caudolateral corner; caudal process of first pleopod much reduced in size.

Allotypic Female.—Differs from the male, form I, chiefly in the reduced size of the chela and shorter fingers; cephalic section of telson with six spines in the caudosinistral and five in the caudodextral corners.

Annulus ventralis subovate with the greatest length in the transverse axis; cephalic margin cleft in middle; sinus originates on midcephalic margin, runs caudad for a short distance, then turns dextrad where it makes a hairpin turn, crosses the midventral line making another hairpin turn across the midventral line where it turns caudosinistrad and terminates just before cutting the midcaudal margin of the annulus.

Measurements.—Male (form I) Holotype: carapace, height 1.41, width 1.39, length 3.25 cm.; areola, width .12, length 1.09 cm.; rostrum, width .54, length .93 cm.; abdomen, length 3.40 cm.; right chela, length of inner margin of palm 1.06, width of palm .87, length of outer margin of hand 2.71, length of movable finger 1.47 cm. Female Allotype: carapace, height 1.58, width 1.64, length 3.41 cm.; areola, width .15, length 1.13 cm.; rostrum, width .53, length .88 cm.; abdomen, length 3.57 cm.; right chela, length of inner margin of palm .70, width of palm .68, length of outer margin of hand 1.92, length of movable finger 1.11 cm.

Type Locality.—At the intersection of State Highway 41 and U. S. Highway 90 at Milligan, Okaloosa County, Florida. Here the specimens were taken from flowing water in a roadside ditch and from simple burrows along its banks.

Disposition of Types.—The holotypic male, the allotypic female, and a second form male paratype are deposited in the United States National Museum. Of the remaining paratypes one male (form I), one male (form II), and one female are deposited in the Museum of Comparative Zoology; one male (form I), one male (form II), and one female in the University of Michigan Museum of Zoology. Ten males (form I), 22 males (form II), 45 females, four immature males, and two immature females are in my personal collection at the University of Florida.

Remarks.—Variation in this species is very marked. The rostra of specimens from Okaloosa County have a broad base with converging margins while those of specimens west of Pensacola have a rostrum which is very narrow at the base and margins which are almost parallel. In one of the specimens taken from north of Olive, Escambia County, Florida, the acumen is almost spiniform, nearly one-third as long as the entire rostrum, and the angles at its base are very acute. In most of the specimens the acumen is not so long nor are the angles at the base so sharp. Considerable variation is also seen in the areola which in the western part of the range is shorter in relation to the cephalic region. The spines on the rostrum, the postorbital ridges, and the carapace are more attentuate in the specimens from Escambia County. Thus specimens from Okaloosa and Escambia counties may easily be distinguished. The first pleopod of the male and the annulus ventralis are remarkably constant.

P. okaloosae very clearly has its closest affinities with *P. clarkii* and *P. paeninsulanus*, but many more detailed locality records and a much larger series of specimens are required for a definite interpretation of its place in this group.

Specimens Examined.—I have a total of 137 specimens of *P. okaloosae* in my collection, 126 of which were collected in Okaloosa, Santa Rosa, and Escambia counties, Florida. The remaining 11 specimens were taken in Escambia County, Alabama. Specimens were collected only during the months of April, May, June, and October; first form males were found in all except October.

Geographical and Ecological Distribution.—This species is at present known from three counties in Florida and one in Alabama. The known range has an east-west extent of about 75 miles and a north-south extent of about 40 miles. Its eastern boundary seems to be marked by the high, well drained sand ridges in Walton County. On the south the range is not known to extend beyond the Yellow River drainage, and the western limit is marked by the tributaries of the Perdido River. There has been insufficient collecting in Alabama to even approximate the northern limit of the range.

Even within the known range this species is poorly known, and much additional field work will be needed to determine the actual range and to make clear its relationships to other forms existing within these limits.

In Escambia County, Florida, *P. okaloosae* has been collected along with *P. blandingii acutus*, *P. versutus*, and *P. spiculifer*; in Okaloosa County with *P. spiculifer* and *P. rathbunae*; in Santa Rosa County with *P. bivittatus* and *C. species incertis*. In Alabama *C. byersi* was taken from burrows near the ditch and swampy area in which *okaloosae* was found.

This species seems to be essentially a small stream inhabitant, although it is by no means confined to such situations. Specimens have been taken from an open roadside ditch, from the backwaters of a river, in small flatwoods pools joining a creek, and once in burrows. Even the creeks from which it has been taken were markedly varied. Some of them were so choked with vegetation that a dip net could scarcely be pushed through the water; others were sand bottomed, and here the crayfish were found under stones or boards in midstream. When this species occurs in streams it usually inhabits the quieter reaches. In one instance (at Riverview, near Pensacola) the crayfish were taken less than a hundred yards from the bay, in the mouth of a very small creek.

The burrows of this species, like those of *P. paeninsulanus*, are always simple, having a single, almost vertical passage.

Procambarus paeninsulanus (Faxon)

Plate VII, Figs. 106-110; Map 6

Cambarus clarkii paeninsulanus Faxon 1914, Mem. Mus. Comp. Zool. 40 (8): 369.
Cambarus clarkii paeninsulanus Faxon 1914: 414; Hobbs 1937: 154; Hobbs (in press-a).
Procambarus clarkii paeninsulanus Hobbs (in press-c).
Procambarus paeninsulanus Hobbs and Marchand (in press).

Diagnosis.—Rostrum with or without lateral spines; areola narrow [somewhat variable in width and length (length varies from 22-36% of carapace)]; male with hooks on ischiopodites of third and fourth pereiopods; postorbital ridges with or without spines; one spine usually present on sides of carapace. First left pleopod of first form male with a strongly developed shoulder on the cephalic surface; shoulder on right is bent mesiocaudad to lie along cephalomesial surface; tip terminating in four distinct parts; mesial process spiniform and directed caudodistad; cephalic process blade-like and usually acute, directed caudodistad, not rounded as in the other three species of the group; caudal process subspatulate and applied to caudolateral surface of central projection, partially hiding its centrocaudal process; mesial process compressed laterally, acute, and directed caudodistad. Annulus ventralis flattened, subovate with the greatest length in the transverse axis; sinus originates near cephalomedian margin, extends caudad for a short distance, turns sinistrad, makes a hairpin turn dextrad crossing the midventral line where it makes another similar turn back to the midventral line along which it proceeds to caudal margin of annulus; mirrored images of the sinus patterns are common.

Remarks.—Much collecting in the panhandle of Florida in search of the intergrades of *clarkii* and *paeninsulanus* has been futile. Furthermore, since the discovery of *P. okaloosae*, a very closely allied species, in Okaloosa and Escambia counties, Florida, between the ranges of *clarkii clarkii* and *clarkii paeninsulanus*, I strongly doubt that *paeninsulanus* is a subspecies of *clarkii*. A distinct break, at least in Florida, has been observed between the ranges of *okaloosae* and *paeninsulanus* in the Walton County region, and I have never taken *clarkii* from this state. Although it is possible that *okaloosae* will prove to be a subspecies of *clarkii*, I cannot believe that intergrading forms will be found between *clarkii* and *paeninsulanus*, and I have therefore proposed that the crayfish referred to *clarkii paeninsulanus* be accorded full specific status as *Procambarus paeninsulanus* until actual data is at hand to prove it a subspecies. There is some indication that there may be intergrades *troglodytes* x *paeninsulanus* in the region just south of the Altamaha River in Georgia, but the evidence is too meager to be convincing.

P. paeninsulanus exhibits considerable variation, and at least some of these variations are correlated with definite areas within the total range.

West of the Apalachicola the length of the areola is from 23 to 33% of the entire length of the carapace; in Washington County, 22-29%; in Holmes, 27-33%; and in Jackson, 27-31%. In the region between the Apalachicola and Suwannee rivers there is a range of 27-34% while east of the Suwannee it is from 28-36%. In specimens from Alachua County it is from 28-36%; in the specimens collected east of the St. Johns River the length of the areola is 33-36% of the entire length of the carapace.

The rostrum varies considerably; in some specimens it is broad and short with converging margins, with or without lateral spines; in others it is narrower with the margins almost parallel and with very strong lateral spines. The acumen may be almost spiculiform or very broad and short. In specimens from Washington County the rostrum is long and narrow, with strong lateral spines and a very long acumen, while elsewhere over the range these two characters are not definitely correlated. The antennal scale, the width of the areola, and the breadth of the carapace also show variation. The postorbital ridges may or may not terminate in spines, and the number of spines in the caudolateral margins of the anterior section of the telson is variable. There is only slight variation in the annulus ventralis, and that in the first pleopod of the male is almost negligible.

At present I am unable to designate any one member of the *clarkii* subgroup—*clarkii, okaloosae,* or *troglodytes*—as its closest relative.

Specimens Examined.—I have a total of 1304 specimens of *P. paeninsulanus* in my collection from Georgia and Florida; of these only 53 are from Georgia. They have been taken from the following counties: GEORGIA—Baker, Camden, Colquitt, Dougherty, Grady, Liberty, Thomas, and Lowndes. FLORIDA—Alachua, Baker, Citrus, Clay, Columbia, Dixie, Duval, Gadsden, Flagler, Gulf, Hamilton, Hillsborough, Holmes, Jackson, Jefferson, Leon, Levy, Liberty, Madison, Marion, Nassau, Putnam, St. Johns, Taylor, Union, Wakulla, Walton, and Washington.

SEASONAL DATA

	Jan.	Feb.	Mar.	Apr.	May	June	July	Aug.	Sept.	Oct.	Nov.	Dec.
♂ I		12	4	45	19		2	43	39	36	6	
♂ II	7	56	18	47	27	5	2		34	21	15	8
♀	11	84	28	71	34	10	7	53	99	50	14	5
♀ (eggs)			1		1			7	20	13	2	
♀ (young)								1	1	2	2	
♂ (immature)	5	25	27	35	2		1	3	21	14	7	2
♀ (immature)	8	27	42	50	9		2	5	13	29	12	1

Geographical and Ecological Distribution.—*P. paeninsulanus* occupies a relatively large area in Florida. Its range extends from the Atlantic westward into the drainage system of the Choctawhatchee River. Its western boundary is marked by the Norfolk-Greenville and Norfolk-Orangeburg Areas (Henderson 1939) in Holmes and Walton counties. The southern boundary of its range in the panhandle probably terminates in the salt marshes along the gulf while in the peninsula it extends into Hillsborough, Marion, Putnam, and Flagler counties. In this section there is no perceptible barrier to restrict this species, and whatever is serving to delimit its range does not affect *fallax* and *alleni*. Collecting in the southern part of Georgia and Alabama, although too inadequate to definitely fix the northern boundary, has been sufficient to permit the approximation indicated on Map 6. It is noteworthy that while *fallax* is not affected by what seems to be the southern limit of the range of *paeninsulanus*, the latter extends beyond the barrier separating *fallax* and *leonensis* as well as that separating *leonensis* and *pycnogonopodus*. It is probable, however, that *paeninsulanus* has avoided these barriers, because its range extends farther to the north, in Georgia, than the northern limits of these barriers.

In Alachua County *P. paeninsulanus* was collected with *P. fallax, seminolae*, and *spiculifer*; in Clay County with *pictus*; in Duval County with *advena* and *seminolae*; in Flagler County with *alleni, fallax*, and *pubischelae*[10]; in Gadsden County with *spiculifer*; in Hamilton County with *seminolae*; in Holmes County with *pycnogonopodus, spiculifer*, and *O. clypeata*; in Jefferson County with *P. leonensis* and *spiculifer*; in Leon County with *spiculifer*; in Levy County with *alleni* and *kilbyi*; in Liberty County with *spiculifer*; in Madison County with *leonensis* and *kilbyi*; in Marion County with *fallax*; in Nassau County with *fallax*; in Putnam County with *fallax*; in Taylor County with *kilbyi* and *leonensis*; and in Washington County with *hubbelli* and *pycnogonopodus*.

P. paeninsulanus, like *fallax*, is nearly ubiquitous, and although it does not seem to inhabit many of the lakes and ponds within the range of *fallax*, it is found to be abundant in the littoral of such habitats in other regions. It has been collected in ponds of all types, lakes, roadside ditches and excavations, sand bottomed and flatwoods streams, rivers, small springs, and burrows.

In situations with an abundant vegetation specimens of *paeninsulanus* are found hiding around the roots, but when the habitat is barren of plant life they hide in piles of debris or in many instances dig shallow burrows into the bank. The mouths of such burrows open below the surface of the water, and the tunnel extends about a foot into the bank.

Although I have seen this species wandering about among the dead leaves at midday in small shaded springs, like most of the crayfish it is more active at night. In most of the sand bottomed creeks in the Gainesville region *paeninsulanus* occurs in large numbers, and at night, with the aid of a headlight, one

---

[10] Determined on the basis of a female specimen which may represent a new species.

may see scores of them in open water around a pile of debris. Frequently I have seen specimens climb from the water onto the bank and remain for a long period before returning to the water.

In at least two instances *paeninsulanus* has been known to wander from the body of water where it lived. On a cloudy day in February a specimen was taken from the middle of a road several hundred yards from the nearest body of water. At another time one was collected on high, well drained land about a quarter of a mile from the nearest body of water.

This species is adept in burrowing, and in a seepage area along the edge of a small fluctuating pond (north of Alachua, Alachua County) numerous burrows of *paeninsulanus* are crowded together in an acre plot. The pond is shallow and rises and falls considerably with the season. In wet weather it often covers the entire area, flooding the burrows of the crayfish, but most of the year it is not so extensive. Time and again I have endeavored to catch specimens with a dip net from the vegetation in the pond proper but have always failed. In the wide exposed margin, however, one may dig a hundred specimens in an hour. The simply constructed burrows are in water-soaked muck which is overgrown with grass and hydrophytic plants, and the digging for crayfish is not at all difficult. The burrows are seldom more than two feet deep and usually consist of a vertical shaft with one or two side passages. During dry seasons the water table in this area is from four to twelve inches below the surface. The chimneys usually consist of a mound of the discarded muck and have no particular shape.

It seems probable that in many instances copulation occurs in the burrow, that the pair separates (there is evidence that it is the male which leaves the burrow) and that the female lays her eggs and remains in the burrow for a considerable period after they have hatched. First form males and females have been found together in burrows all over the range, and practically all of the females carrying eggs or young were found in burrows.

The Evermanni Subgroup

A separate *evermanni* subgroup is necessitated to receive the very disjunct *Procambarus evermanni*.

Procambarus evermanni (Faxon)

Plate VII, Figs. 111-115; Map 7

Cambarus evermanni Faxon 1890, Proc. U. S. Nat. Mus. 12 (785): 620-621.
Cambarus evermanni Ortmann 1902: 277; Ortmann 1905a: 98, 102, 105; Ortmann 1906: 18; Harris 1903: 58, 97, 144, 151; Faxon 1914: 414.
Procambarus evermanni Hobbs (in press-c).

Diagnosis.—Margins of rostrum interrupted, usually with small lateral spines; areola relatively broad with about three punctations in narrowest part; male with hooks on ischiopodites of third and fourth pereiopods; hooks prominent but simple; chela of first form male never bearded along inner margin of palm but bears an irregular row of about seven tubercles; postorbital ridges terminating cephalad in small spines or tubercles; no lateral spines present on sides of carapace. First pleopod reaching base of third pereiopod; tip terminating in four distinct parts; (the Mississippi specimens lack the cephalic process;) cephalic process, if present, rather small and spiniform and projects caudodistad to partially hood the central projection; caudal process well developed, long, and somewhat spatulate (in lateral view), and is closely applied to lateral margin of central projection, practically obscuring the centrocaudal process; central projection conspicuous and directed caudodistad; viewed laterally only the centrocephalic process can be seen; no shoulder present on cephalic margin of appendage. Annulus ventralis diamond-shaped with greatest length in the transverse axis; somewhat flattened; sinus, originating on midventral line slightly caudad of cephalic margin, curves caudodextrad, then gently caudosinistrad reaching midventral line about one-fourth of length of annulus from caudal margin where it turns caudad to midcaudal margin.

Remarks.—The Florida specimens of *Procambarus evermanni* seem to be fairly constant, but the areolae of the Escambia County specimens are broader than in those from Santa Rosa and Okaloosa counties. All of these specimens have small lateral spines on the rostrum, and in addition, some have small lateral spines on the carapace. The Mississippi specimens of *evermanni* show several striking differences from those from Florida. The rostrum does not bear lateral spines; the postorbital ridges terminate in tubercles rather than spines; the cephalic process of the first pleopod is absent, and the central projection appears to be somewhat larger; there is also some difference in the caudal process of this same appendage. Perhaps this group of specimens from Mississippi should be regarded as a separate species, but since the range of *evermanni* is so poorly known, it will be necessary to wait until thorough collecting has been done in Baldwin and Mobile counties, Alabama, to determine whether there is an actual break between the two groups.

Procambarus evermanni is one of the least known of the North American crayfishes, and I can add but little to the quotation that follows. "The other species, *C. evermanni* and *barbatus*, are known from scattered localities in Georgia, western Florida and Mississippi, and their distribution needs further investigation; . . . ." (Ortmann 1905a: 105). Certainly the Georgia record needs confirmation. (It has already been pointed out that *P. barbatus* is found only in Georgia and South Carolina.)

Most of the specimens in the museums attributed to *P. evermanni* are *P. alleni*; in fact the only specimens of *evermanni* I have seen outside of my own collection were the type and a few specimens at the Museum of Comparative Zoology, which were taken from Escambia County, Florida. It is unfortunate that the mesial process of the first pleopod of the type is broken, and probably this is the reason that other species have been confused with it.

P. evermanni has been considered by authors to have close affinities with

P. wiegmanni, P. fallax, P. barbatus, and *P. alleni.* Of these *fallax* probably is most nearly allied to *evermanni,* but even it does not seem to be so closely related as are *paeninsulanus, hayi,* and *blandingii acutus.* Faxon (1890: 620) referred *evermanni* to the group of *blandingii* but then stated that "It is nearly related to *C. alleni* Fax.". As has been pointed out above, *alleni* should not be placed in the *blandingii* group (its affinities seem to be with *gracilis, simulans, hagenianus,* etc.). I believe that *evermanni* is correctly referred to the *blandingii* section, but that its closest affinities are not those pointed out by previous authors. The first pleopod of the male is most like that of *paeninsulanus* and is also similar to *hayi* and *blandingii acutus.* The pleopod of the second form male in the latter when compared with *evermanni* definitely indicates close affinites. The pleopods of all four of these species (*evermanni, paeninsulanus, hayi,* and *blandingii acutus*) have a similarly well developed caudal process and a comparable arrangement of the other terminal processes. In *evermanni, hayi,* and *blandingii acutus* the sternum of the female, just cephalad of the annulus ventralis, is tuberculate, and the tubercles are large and extend caudad to overhang a part of the annulus. In *evermanni* this region of the sternum as well as the annulus ventralis approaches that of *hayi* more closely than it does those of the others. Other characters—the rostrum, areola, chelae, etc.—support this relationship, and the ecological and geographical data, so far as known, do not conflict with it.

Specimens Examined.—I have a total of 80 specimens in my collection, and I have examined the type specimens and several additional specimens in the Museum of Comparative Zoology. My specimens were collected in the following localities: FLORIDA—Okaloosa County—5.4 miles west of Fort Walton [U. S. Highway 98]; 1.6 miles west of Fort Walton [U. S. Highway 98]; 11.2 miles west of Fort Walton [U. S. Highway 98]. Santa Rosa County— 14.8 miles west of Fort Walton [U. S. Highway 98]; 13.9 miles west of the Okaloosa County line [U. S. Highway 98]; .5 miles east of Navarre [U. S. Highway 98]; 5 miles north of intersection of U. S. Highway 98 and State Highway 10. MISSISSIPPI—Jackson County—5.3 miles west of Grand Bay [U. S. Highway 90]. Specimens have been collected only in the months of April and June, and first form males were taken in both months.

Geographical and Ecological Distribution.—The range of this species in Florida includes the Escambia River drainage system and the southern parts of Santa Rosa and Okaloosa counties. Further collecting will probably show that the range is more extensive (perhaps extending into the drainages of the Yellow and Blackwater rivers); however, I do not believe that *evermanni* will be found farther east in Florida than the eastern part of Walton County. The only locality known for Alabama is in Escambia County near Flomaton (almost on the Florida state line). In Mississippi I have collected it from Jackson County.

In Okaloosa County *evermanni* was taken along with *versutus* in one locality, and in Santa Rosa County *byersi* was found burrowing on the banks of a small ditch which joined the stream in which *evermanni* occurs.

All of my Florida specimens were taken from small creeks in the coastal lowlands of Walton and Santa Rosa counties. Near Fort Walton, Okaloosa County, specimens were obtained from a leaf drift (about 100 yards from Santa Rosa Sound) in a small, clear, sand bottomed creek. Two of the collections made in Santa Rosa County were in small, sluggish, lowland streams. In both streams the crayfish were found in accumulations of dead leaves and twigs in the still portions of the stream. In some places these accumulations were as much as two and one-half feet deep and often contained several crayfish. In only two instances did I take this species from open water; both of these were from sand bottomed creeks with dense aquatic vegetation, and it is probable that in pushing my dip net through the vegetation I had frightened them from their hiding places so that they retreated to open sandy areas.

Map 7.—Distribution of the Evermanni and Fallax Subgroups.

The Fallax Subgroup

Diagnosis.—The first pleopod of the first form male terminates in three (sometimes four) distinct, though comparatively small parts; mesial process either spiculiform or blade-like; cephalic process, viewed mesially, a slender spiniform process directed caudad; caudal process often absent, but when present a small projection from caudal side of ridge which flanks the central projection; central projection a corneous outgrowth from the center of the distal truncate terminal; it may be compressed laterally (in *fallax* and *pycnogonopodus*) or obliquely (in *leonensis*) and may be directed either caudodistad (in *fallax*) or distad (in *leonensis* and *pycnogonopodus*); it may be conspicuous (in *fallax*) or decidedly reduced (in *pycnogonopodus*). The areola is relatively long and narrow; lateral spines are usually present on the rostrum, and hooks are present on the ischiopodites of both the third and fourth pereiopods.

The *fallax* subgroup comprises three nearly ubiquitous species—*fallax, leonensis,* and *pycnogonopodus*, at least one of which is to be found in almost any aquatic situation in Florida, from the drainage system of the Choctawhatchee River to the Atlantic, and southward to the region of Lake Okeechobee. (Map 7).

Procambarus fallax (Hagen)

Plate VII, Figs. 116-120; Map 7

Cambarus fallax Hagen 1870, Mem. Mus. Comp. Zool. 3: 45-46, Pl. I, figs. 103-105.
Cambarus fallax Hagen 1870: 97, 101, 107; Faxon 1884: 136; Faxon 1885a: 17, 19, 23, 24, 29, 157, 173, Pl. II, fig. 4; Faxon 1885b: 357; Faxon 1890: 621; Faxon 1898: 644; Ortmann 1902: 277; Harris 1903: 58, 70, 97, 143, 152, 166; Ortmann 1905a: 102, 105; Faxon 1914: 368, 413; Creaser 1934: 4; Hobbs 1937: 154.
Procambarus fallax Hobbs (in press-c); Hobbs (in press-b).

Diagnosis.—Rostrum usually with lateral spines; areola narrow, with two or three rows of punctations in narrowest part; male with hooks on ischiopodites on third and fourth pereiopods; postorbital ridges terminating cephalad in spines; a lateral spine present on each side of carapace. First pleopod of first form male without a shoulder on cephalic section and reaching coxopodite of second or third pereiopod; tip terminating in four parts; mesial process slender and blade-like and directed laterodistad; cephalic process spiniform and directed caudodistad but closely applied to cephalodistal portion of appendage; caudal process small and inconspicuous and situated laterad of the central projection; central projection acute, corneous, compressed laterally, and directed caudodistad. Annulus ventralis somewhat bell-shaped in outline, narrow cephalad, and swollen in caudal portion; high lateral ridges not uncommon along cephalolateral surfaces; sinus originates on or near cephalomedian margin and follows midventral line half the length of annulus where it turns sharply sinistrad, makes a hairpin turn and crosses the midventral line, once more makes a hairpin turn back to midventral line, turns caudad, and terminates on or near midcaudal margin of annulus; mirrored images of this sinus pattern are common.

Remarks.—*P. fallax* is an extremely variable species, but there seem to be no variants which can be correlated with definite regions; specimens from a single locality show variations as diverse as specimens taken at opposite limits of the range.

The most outstanding variations are seen in the rostrum, chelae, and areola. The rostrum varies tremendously in length and usually bears two marginal spines which are moderately developed; however, I have specimens with additional spines, some without a trace of a spine and only interrupted margins, and some with the usual two spines strongly developed. The acumen is often long and spiculiform, although occasionally it is short and approaches in shape an equilateral triangle. The chelae vary in proportion and in the number of tubercles along the inner margin of the palm. The areola varies in width and is often depressed, but in some specimens stands above the rest of the carapace so that it appears as a ridge. I am able to recognize specimens from certain regions on color patterns, but this is not always reliable, for there is some variation even here.

P. fallax is most nearly related to *pycnogonopodus* and *leonensis*, and it also seems to have affinities with *evermanni*.

Specimens Examined.—I have examined a total of 1844 specimens of *P. fallax*; all except 12 (which are from Echols County, Georgia) were collected in Florida from the following counties: Alachua, Baker, Citrus, Clay, Columbia, DeSoto, Flagler, Gilchrist, Hamilton, Hardee, Hendry, Hernando, Highlands, Hillsborough, Indian River, Lake, Levy, Marion, Nassau, Orange, Palm Beach, Pasco, Pinellas, Polk, Putnam, Seminole, St. Johns, Sumter, Suwannee, Taylor, Union, and Volusia.

SEASONAL DATA

	Jan.	Feb.	Mar.	Apr.	May	June	July	Aug.	Sept.	Oct.	Nov.	Dec.
♂ I	26	36	60	28	4	14	6	5	24	23	17	3
♂ II	12	73	49	22	2	8	8		42	64	64	12
♀	34	141	85	56	17	20	11	4	73	89	74	16
♀ (eggs)	1	6	1	6	3	8		2	10	3	2	1
♀ (young)				2		2						
♂ (immature)	16	42	20	7	3	1	38	3	36	38	24	23
♀ (immature)	27	48	25	20	14	2	30	7	32	51	34	16

Geographical and Ecological Distribution.—*Procambarus fallax* occupies a range comparable in size to that of *alleni* and *paeninsulanus* and is one of the most common species in the north-central part of the peninsula. Its range is marked on the west by the extensive area of Blanton-Norfolk soils

(Henderson 1939) in Hamilton, Suwannee, Madison, Lafayette, Gilchrist, Levy, and Dixie counties and farther south by the Gulf of Mexico. On the north there is no perceptible barrier, and perhaps *fallax* is more widespread in the southeastern part of Georgia than my locality records show. (It seems that in the Okefenokee region this species is replaced by *seminolae*.) The eastern boundary is marked by the salt marshes of the coast, while the southern limits of its range, like the northern, are indefinite, seeming to end in DeSoto, Highlands, and Palm Beach counties without any apparent physical barrier.

The following is a summary of the recorded distribution of this species:

FLORIDA—(Hagen 1870); Lake Jessup, Seminole County, Indian River, Brevard (?) County, near Titusville, Brevard County, St. Johns River, Putnam (?) County (Faxon 1885b); Horse Landing, Putnam County, Hawkinsville, (?) County, Orange Bluff, Nassau County, Blue Spring, (?) County, Lake Jessup, Seminole County, Magnolia, (?) County, Indian River, (?) County, Titusville, Brevard County (Faxon 1885a); Eustis, Lake County, Gainesville, Alachua County (Faxon 1898); Auburndale, Polk County, Kissimmee River, Osceola County, Lake Monroe, Seminole County, St. Johns River at Palatka, Putnam County, St. Johns River at Beecher Point, Putnam County (Faxon 1914); Gainesville Region, Alachua County (Hobbs 1937).

In Alachua County this species was collected with *paeninsulanus, schmitti*, and *spiculifer;* in Citrus County with *alleni;* in Columbia County with *seminolae;* in Flagler County with *paeninsulanus, alleni*, and *pubischelae* (?); in Gilchrist County with *spiculifer;* in Hamilton County with *seminolae;* in Hillsborough County with *alleni;* in Marion County with *paeninsulanus;* in Nassau County with *paeninsulanus* and *spiculifer;* in Putnam County with *paeninsulanus;* in Seminole County with *alleni;* in Sumter County with *alleni;* in Union County with *seminolae, pubischelae*, and *paeninsulanus*.

P. fallax, unlike many of the Florida species, is not restricted to one particular type of habitat, and it shows little, if any, correlation with any one type of lake or pond, so long as there is sufficient vegetation to afford hiding places and there is water most of the year.

Practically all of the streams within the range of *fallax* seem to be inhabited by it, in at least some of their reaches. In larger calcareous streams and many of the spring runs *P. fallax* is abundant. Even in some of the helocrene springs small specimens have been taken from the leaves which have fallen into the rill courses. Many of the acid flatwoods streams are inhabited by *fallax*, and in several instances this species has been taken from sand bottomed creeks.

Although *P. fallax* inhabits both lotic and lenitic situations in the northern and central parts of its range, it becomes more restricted to lotic habitats as it approaches the southern limits. It is most often found in the quieter reaches of a stream, or if there is abundant vegetation, as is the case of many of the spring runs, many of the crayfish are found hiding among the plants. In general there seems to be a high degree of correlation between the abundance of *fallax* and the amount of vegetation. In some of the sand bottomed creeks with a sparse flora, *fallax* is absent, but where the creek becomes more sluggish and plants become more abundant, *fallax* is usually common. Never have I taken this species from a sand bottomed stream where there was no vegetation.

This species probably does not burrow by preference but when the water table is lowered in dry seasons, *fallax* usually constructs a simple burrow with only one passage slightly slanting from the vertical; at the bottom, which may or may not be below the permanent water table, a slightly enlarged chamber occurs; it is here the crayfish is always found when it is dug from the burrow.

Procambarus alleni and *fallax* have been found together in roadside ditches which drain into a stream. When the water is backed up into the ditch the two species are taken in about equal numbers; however, when there is a period of drought and the ditch becomes dry many more burrows of *alleni* are found, and the adjoining streams are thickly populated by *fallax*.

Alachua County, more thoroughly collected than others of the state, has a tremendous *fallax* population. Here this species has been collected from lakes, ponds, flatwoods, roadside excavations, ditches, springs, in the Santa Fe River, and in several instances during wet weather crawling around in fields and hammocks. During dry seasons it may be taken in large numbers from burrows in pond basins or ditches.

In the sand-scrub lakes of Marion County *P. fallax* seems to be the sole crayfish occupant. In July 1938 Professor T. H. Hubbell and Mr. J. J. Friauf were lighting for insects in this region. The light was about 200 yards from Niggertown Lake, the nearest body of water, in a mesic hammock of live oak and sweet gum. A male of *P. fallax* was picked up at the light. In Alachua County *fallax* was taken from a hole, dug to catch frogs, about 75 yards from Lake Newnan. At the edge of one of the deep sink holes I observed one specimen crawling around on the bank. In Marion County *fallax* is common in the lakes, creeks (particularly the low, sluggish ones), swamp ponds, the Withlacoochee River, and its flood plains where it digs simple burrows in low moist places. In the river it was taken from the roots of hyacinths and water lettuce.

Probably because of the tremendous amount of gill space, this species can live in bodies of water in which the oxygen content is very low, and I have taken individuals in sphagnum bogs where the pH of the water was less than 5. Specimens found in this last type of situation always have a very thin, somewhat pliable carapace, very different from the hard carapace of the specimens of *fallax* found in calcareous streams.

Procambarus leonensis Hobbs
Plate VIII, Figs. 121-125; Map 7

Procambarus leonensis Hobbs (in press-b), Proc. Fla. Acad. Sci.

Diagnosis.—Rostrum with small lateral spines or tubercles; areola moderately broad with two rows of punctations in narrowest part; male with hooks on ischiopodites of third and fourth pereiopods; postorbital ridges terminating cephalad in spines or tubercles; lateral spines present on carapace. First pleopod of male with no shoulder present on cephalic surface, reaching coxopodite of third pereiopod; tip terminating in four parts; mesial process subspiculiform and directed caudolaterad; cephalic process also subspiculiform and extends from cephalic portion of tip in a caudodistal and somewhat lateral direction; caudal process vestigial, slightly curved, and lies at the base of centrocaudal process of central projection; central projection small, corneous, subspatulate, and extends distad from center of outer part of appendage. Annulus ventralis subtriangular; caudal margin symmetrically bilobed; sinus originates along midventral line near cephalic margin, extends gently caudosinistrad, then caudodextrad to midventral line, and finally caudad almost to caudal margin of annulus, terminating on caudal surface of a small mound.

Remarks.—Little variation has been noted in this species except that some of the specimens from the vicinity of Lamont seem to have a relatively broader rostrum. The areola seems to be slightly shorter in the western part of the range than elsewhere. As in most species, there are a few specimens with additional rostral spines; one of these has five lateral spines on the acumen of the rostrum.

P. leonensis along with *fallax* and *pycnogonopodus* shows certain relationship to the members of the *clarkii* subgroup as well as to the *pictus* group and *blandingii* subgroup, but probably *evermanni* is its closest relative outside of the *fallax* subgroup.

Specimens Examined.—I have examined 382 specimens of *P. leonensis*, from the following counties in Florida: Franklin, Gadsden, Jefferson, Lafayette, Leon, Liberty, Madison, Taylor, and Wakulla.

SEASONAL DATA

	Jan.	Feb.	Mar.	Apr.	May	June	July	Aug.	Sept.	Oct.	Nov.	Dec.
♂ I	3			4	4	5		1			9	1
♂ II		7	5	5	5	2					10	1
♀		11	4	8	7	4		3			15	2
♀ (eggs)				2	2			1				
♀ (young)												
♂ (immature)		20	2	1	46	1					33	
♀ (immature)		51	8	2	68						27	2

Geographical and Ecological Distribution.—This crayfish appears to be specifically distinct in spite of its evidently close relationship to *P. fallax*. These two are not only morphologically similar but also their ranges are adjacent—separated in some places by only a few miles. It is possible that somewhere in this strip of territory separating the two species, intergrading specimens will be found when the region has been intensively collected. (The most likely region for intergradation lies in the southern part of Georgia in the region of the Aucilla and Withlacoochee rivers.) If intergrades occur between *leonensis* and *pycnogonopodus* they probably will be found in Franklin and Gulf counties. *P. leonensis* and *fallax* are undoubtedly derived from the same stock, and the differences between them are very slight—so slight in fact that some of the females are almost indistinguishable.

P. leonensis is known from scattered localities in all of the counties lying between the Apalachicola and Suwannee rivers, except Dixie. The eastern and western boundaries of the range of this species are not the rivers themselves. The eastern boundary is probably the divide between the drainage systems of the Suwannee and the Econfina, Fenholloway, and Aucilla rivers. This is a strip of land with very few streams and only an occasional lake or pond, and extends through the eastern part of Madison, and southward through Lafayette and Dixie counties. The western boundary of the range seems to be in the high, dry strip of territory east of the Apalachicola River. In Madison County *P. leonensis* was collected with *paeninsulanus* and *kilbyi*; in Gadsden County with *spiculifer*; in Jefferson County with *paeninsulanus* and *kilbyi*; in Leon County with *spiculifer* and *paeninsulanus*; in Liberty County with *spiculifer*; and in Taylor County with *paeninsulanus*.

Like *fallax*, *leonensis* occupies many types of habitats. It has been collected from swamp ponds, acid creeks, clumps of sphagnum, roadside ditches, sand bottomed streams, lakes, and burrows. Most of the mature specimens in my collection were taken from burrows; however, too little is known of the habits of this species to assume that unlike *fallax*, it burrows by preference when adult. More probably it is purely a coincidence that most of my collections were made either in dry weather or shortly thereafter. I know that this is the case for most of my collecting in Taylor County. In Franklin County this species was taken from a sand bottomed stream where *Hydrocotyl*, *Persecaria*, *Ceratophyllum*, and *Utricularia* were growing along the margins, and most of the specimens were taken among the roots of this vegetation. In Gadsden County *leonensis* was taken from a roadside ditch adjoining a small, clear, moderately swift stream. Some of the specimens were found in the open ditch, but several were taken from flooded burrows in its bottom. In Jefferson County this species was collected from burrows in the edge of a temporary pond. These burrows were simple, and although only about a foot deep penetrated the water table. In Lafayette County specimens were dug from burrows in a roadside ditch which usually contains water. The burrows here penetrated to a depth of about one foot but did not reach the water table. From Leon County *leonensis* was taken from acid swamp pools, from burrows around open mud bottomed ponds, and from lakes and sloughs. Several of the lakes in Leon County are sporadically drained by subterranean outlets; at such times the entire lake goes dry except for a few small pools, and the crayfish may be found burrowing in the mud bottoms or may be taken in seines dragged through the pools. Judging from the relative abundance of burrows and the number of crayfish taken in the seines, this species must be common in these large lakes. In a small swamp stream near Tallahassee *leonensis* was taken from the edge of the water among *Eleocharis*, *Ludwigia*, and *Sphagnum*. In

Liberty County it was collected from a sand bottomed creek. In three localities in Madison County this species was taken from small streams which were choked with vegetation. Two of these streams were flowing rather rapidly and had an abundant growth of *Vallisneria* and other aquatics. The other stream was rather sluggish and was continuous with a swamp pond supporting an abundant growth of *Castalia*. From Taylor County *leonensis* was taken in a shallow pool in a Cypress-Gum swamp, from roadside ditches, and from pools in the bed of a small, stagnant, swamp stream.

Procambarus pycnogonopodus Hobbs

Plate VIII, Figs. 126-130; Map 7

Procambarus pycnogonopodus Hobbs (in press-b), Proc. Fla. Acad. Sci.

Diagnosis.—Rostrum with small lateral spines or tubercles, or margins only slightly interrupted; areola of moderate width with two or three punctations in narrowest part; male with hooks on ischiopodites of third and fourth pereiopods; postorbital ridges terminating cephalad in spines, tubercles, or they may be truncate; lateral spines may or may not be present on sides of carapace. First pleopod of male with no shoulder present on cephalic surface, reaching coxopodite of third pereiopod; tip terminating in three parts; mesial process subspiculiform and extends in a distal caudolateral direction; cephalic process subspiculiform, arises from cephalomesial part of appendage, and directed distad or sharply recurved caudad; central projection very small, blade-like, and extends distad from the center of terminal portion of the outer part. Annulus ventralis broadly bell-shaped; caudal margin shallowly cleft in middle; sinus originates on midcephalic margin, follows rather closely the midventral line to midlength of annulus, bending gently dextrad, for a short distance caudad, then sinistrad to midventral line, and finally caudad to cut the midcaudal margin.

Remarks.—Some of the specimens have a relatively long narrow rostrum with the sides hardly converging while in others the margins are strongly convergent and the lateral spines are almost microscopic, thus giving the whole structure a triangular appearance. In other specimens the rostrum is very broad at the base with the margins slightly converging and terminating in a very short, blunt acumen. Lateral spines may or may not be present; if present they may or may not be strongly developed. Slight variations have been noted in the relative proportions of the carapace, but they do not seem to be constant for any local population. Other variations also occur in the length and width of the areola, the various spiny parts, and the chelae. The first pleopod of the first form male apparently is the least variable character to be appealed to, and even here the cephalic process varies in shape from straight and spiculiform to curved at a right angle [Hobbs (in press-b)].

P. pycnogonopodus has its closest affinities with the members of the *fallax* subgroup, and in some respects it seems to be more closely related to *fallax* than to *leonensis*.

Specimens Examined.—I have examined a total of 559 specimens, from the following counties in Florida: Bay, Calhoun, Gulf, Holmes, Jackson, Okaloosa, Walton, and Washington.

SEASONAL DATA

	Jan.	Feb.	Mar.	Apr.	May	June	July	Aug.	Sept.	Oct.	Nov.	Dec.
♂ I				23	2	8					7	2
♂ II				50	8	18					23	22
♀				59	8	19					55	29
♀ (eggs)				5	1						3	
♀ (young)												
♂ (immature)				29	2						26	39
♀ (immature)				46	2						35	38

Geographical and Ecological Distribution.—This species, though easily distinguishable from the other members of the *fallax* group, shows very definite morphological relationship with them, and the habitats from which it has been taken are not markedly different from similar situations in which *fallax* and *leonensis* occur. The barrier, if any, that separates *leonensis* and *pycnogonopodus* is in the vicinity of the Apalachicola River. The western limit of the range of *pycnogonopodus* lies in the territory between the drainage systems of the Yellow and Choctawhatchee rivers. Several of the small streams emptying into Choctawhatchee Bay east of Niceville are inhabited by this species. The westernmost locality known is about seven miles east of Destin, Okaloosa County. A line drawn from the point where the Walton-Okaloosa County line meets the Choctawhatchee Bay, through the city of DeFuniak Springs, thence to the point where the Holmes-Walton County line intersects the Alabama state line, approximates the western limit of the range of this species. The southern limit of the range is marked by the salt water of the Gulf of Mexico and its bays. In the east the range terminates in the vicinity of the Apalachicola River in Jackson, Calhoun, and Gulf counties. Since this species is found in the Choctawhatchee drainage system it is not improbable that its range extends into at least the southern part of Alabama.

The actual territory encompassed by the limits of the range of *pycnogonopodus* is considerably smaller than that of *fallax* and slightly smaller than that of *leonensis*.

P. pycnogonopodus has been collected with several of the west Florida species, namely: *C. schmitti, P. latipleurum, apalachicolae, kilbyi, paeninsulanus, hubbelli, spiculifer,* and *versutus*. In Bay County *pycnogonopodus* was taken with *spiculifer, latipleurum,* and *apalachicolae;* in Calhoun and Gulf counties with *kilbyi;* in Holmes County with *paeninsulanus* and *hubbelli;* in Jackson County with *paeninsulanus;* in Walton County with *schmitti, versutus,* and *hubbelli;* in Washington County with *paeninsulanus, hubbelli,* and *spiculifer*.

A habitat of *P. pycnogonopodus* seems to be provided by any body of water, either temporary or permanent, located in a low swampy region. It is commonly found in cypress ponds, bayheads, flatwoods streams, and roadside ditches. In several instances *pycnogonopodus* has been taken from pools in the drying bed of a stream. In normal water stages, it is found in leaf drift, beneath sunken logs in clumps of *Utricularia, Eleocharis, Sagittaria, Vallisneria,* and among the roots of *Pontederia, Orontium,* or *Persicaria*. Individuals seem to be more abundant in bodies of water supporting a vigorous growth of aquatic vegetation.

In only a few instances has this species been taken in a different situation from those listed above. In Walton County, about 16 miles east of Niceville, a small stream from a flatwoods pond flows into the bay. Here several specimens of *pycnogonopodus* were taken from the stream proper, and one was taken about 50 yards from the mouth of this stream in Choctawhatchee Bay, where the water is definitely brackish. Several clear streams also flow into the bay from its northern shore in Walton County. In several of these *pycnogonopodus* was collected within 15 yards of the bay.

The Spiculifer Group

Diagnosis.—The first pleopod of the first form male terminates in three or four parts (the cephalic process is lacking in *spiculifer*) with no hump along the cephalic margin of the appendage; mesial process spiculiform and always prominent; cephalic process when present relatively small and in apposition with and partially hooding the central projection; central projection conspicuously corneous and making up the bulk of the terminal part of the appendage, although it is relatively slenderer; caudal process arises from the base of and caudolaterad of the central projection and is the least conspicuous of the four terminal processes—in *spiculifer* it is in such close proximity with the central projection that it may be mistaken for a part of it. Rostrum bears a well developed lateral spine along either margin and terminates in a long acumen; areola wide and much shorter than cephalic section of carapace; hooks present on ischiopodites of third and fourth pereiopods in male.

Two species, *spiculifer* and *versutus*, are referable to this group, and both of them are confined to lotic situations in Mississippi, Alabama, Georgia, and Florida.

Procambarus spiculifer (LeConte)

Plate VIII, Figs. 131-135; Map 8

Astacus spiculifer LeConte 1856, Proc. Acad. Nat. Sci. Philad. 7: 401.
Cambarus spiculifer Hagen 1870: 9, 10, 31, 33, 48, 49, 50, 52, 97, Pl. I, figs. 59-62, Pl. III, fig. 147; Faxon 1884: 138; Faxon 1885a: 17, 18, 19, 30, 31, 33-34, 158, 173, Pl. II, fig. 5; Faxon 1885b: 358; Ortmann 1902: 277; Harris 1903: 58, 130, 138, 143, 144, 152; Ortmann 1905a: 100, 101, 105, 126, 128; Creaser 1934: 4; Hobbs 1937: 154; Lyle 1938: 76; Hobbs (in press-a).
Procambarus spiculifer Hobbs (in press-c).

Specimens Examined.—I have examined a total of 1038 specimens of *P. spiculifer*, 860 of which were collected in Florida, 152 in Georgia, and 26 in Alabama. The following is a list of localities compiled from the literature. GEORGIA—Athens, Clarke County (Hagen 1870); Milledgeville, Baldwin County; Atlanta, Fulton County; Roswell, Cobb County; Gainesville, Hall County; Etowah River, (?) County (Faxon 1885a). MISSISSIPPI (Lyle 1938).

SEASONAL DATA

	Jan.	Feb.	Mar.	Apr.	May	June	July	Aug.	Sept.	Oct.	Nov.	Dec.
♂ I	3		2	6	21	4	5			4	7	6
♂ II	1	8	34	44	23	49	5	10	1	11	25	14
♀	10	5	34	44	17	50	6	11	2	21	49	12
♀ (eggs)	3		1				1					
♀ (young)												
♂ (immature)		5	14	71	28	26		23		40	49	8
♀ (immature)		4	24	50	12	17		31		41	32	6

Geographical and Ecological Distribution.—*Procambarus spiculifer* has been collected from scattered localities over a large part of Georgia and Florida and from a few isolated localities in the southeastern part of Alabama.

A complete list of states and counties in which *spiculifer* has been recorded or from which I have collected it is given below: ALABAMA—Baldwin, Escambia, Dale, and Pike counties. GEORGIA—Baker, Baldwin, Brooks, Bryan, Clarke, Cobb, Colquitt, Decatur, Dougherty, Early, Fulton, Grady, Greene, Hall, Houston, Jones, Lowndes, Mitchell, Oconee, Roswell, and Sumter counties. MISSISSIPPI—no county cited. FLORIDA—Alachua, Bay, Calhoun, Columbia, Escambia, Gadsden, Gilchrist, Hamilton, Jackson, Jefferson, Leon, Levy, Liberty, Nassau, Okaloosa, Suwannee, Union, Wakulla, Walton, and Washington counties.

Since this species is primarily a stream form, its range will be discussed in terms of the larger drainage systems in which it occurs. In Georgia *P. spiculifer* is known to inhabit at least parts of the drainage systems of the Oconee, Ocmulgee (both of these are tributaries of the Altamaha River), Chattahoochee, Etowah, Flint, Ochlocknee, and Suwannee rivers. It is evident that the northernmost records of *spiculifer* are in the Etowah, Oconee, Ocmulgee, and Chattahoochee rivers, and that the northeastern and eastern limits of its range are indicated by the tributaries of the Oconee and Altamaha rivers. [It has been recorded from neither the Ogeechee nor the Savannah rivers, where *pubescens* occurs, and there is little doubt that these two species exhibit vicariation. (Hesse, Allee, Schmidt 1937:78).] In the northwest the range is probably limited to the tributaries of the Coosa River, and the Alabama River is the westernmost stream from which I have seen specimens.

In Florida *spiculifer* is found in most of the major rivers west and north of the peninsula proper. In the panhandle it is known from the Perdido, Yellow, Choctawhatchee, Chipola, Apalachicola, and Ochlocknee rivers. Farther east it has been collected in the Suwannee, Santa Fe, and tributaries of the St. Marys River, and the most southern localities known are the Suwannee and Santa Fe rivers.

Because of the relatively large list of crayfish associates it should not be supposed that this form occupies a wide range of habitats, for, as will be pointed out below, *spiculifer* is for the most part most exacting in its choice of habitats.

Associated with *spiculifer* in different parts of its range are *pycnogonopodus, versutus, fallax, okaloosae, schmitti, leonensis,* and *paeninsulanus.*

P. spiculifer is the largest of the Florida crayfishes. Typically it occurs in rather clear streams in which the oxygen content is high. (Not only does *spiculifer* have relatively small gill chambers, but laboratory observations indicate that it is unable to survive in a low oxygen concentration.) The size of the stream is not of primary importance—almost any well aerated, lotic situation which the crayfish can reach by suitable highways is likely to be occupied. *P. spiculifer* has never been collected in standing water, and only rarely have I taken it from the still reaches of a stream. Neither has it been recorded from the acid[12], sluggish flatwoods streams even when they were directly connected with larger streams in which *spiculifer* is abundant.

Since the Santa Fe River is somewhat illustrative of the selective habit of this species a brief description of its course is given below. The river originates as a temporary stream draining Santa Fe Lake, which is in the extreme northeastern part of Alachua County. The lake, as well as the upper portion of the river, receives tributaries and surface drainage from the flatwoods in the eastern part of the county, and in this region the water has the dark brown color of acid flatwoods drainage. Some 18 to 20 miles below its source, the Santa Fe, now a small permanent creek, receives the waters of several large springs, and within a few miles loses the characteristics of a flatwoods stream. Eight or ten miles below the first large spring the river disappears into the underlying limestone through which it flows for fully two miles before it emerges to flow again as a surface stream. As a result of passing through the limestone and of receiving additional water from several large springs, the water becomes circumneutral to basic and comparatively cool, although it may retain a slight brownish tinge. In some places the river flows over a bare limestone stratum at a rate of two to three miles an hour. After heavy rains the stream temporarily loses its striking calcareous character, and the water, now dark brown, often floods the surrounding swamps.

Several hundred specimens have been taken at intervals along the course of the river. The greater number has been caught in the riffles below the natural bridge. These riffles range in depth from one to four feet, and large rocks may be seen extending above the water line. The rocky and sandy bottom supports an abundance of submerged vegetation. The flora of these riffles con-

---

[12]Some of the streams inhabited by *spiculifer* have brief periods in which the water is coffee-colored from swamp drainage.

Procambarus versutus (Hagen)

Plate VIII, Figs. 136-140; Map 8

Cambarus versutus Hagen 1870, Mem. Mus. Comp. Zool. 3: 51-52, Pl. I, figs. 55-58, Pl. III, fig. 150.

Cambarus versutus Hagen 1870: 28, 31, 34, 97, 101, 106, 107; Faxon 1884: 138; Faxon 1885a: 17, 18, 19, 31, 33-34, 158, 173; Faxon 1885b: 358; Faxon 1890: 619; Faxon 1898: 646; Ortmann 1902: 277; Harris 1903: 58, 131, 144, 151, 152; Ortmann 1905a: 101, 128; Faxon 1914: 367, 412; Creaser 1934: 4; Lyle 1938: 76.

Procambarus versutus Hobbs (in press-c).

Diagnosis.—Rostrum with lateral spines, a longitudinal carina present on upper surface; areola broad with 8-15 punctations in narrowest part; male with hooks on ischiopodites of third and fourth pereiopods; postorbital ridges terminating cephalad in spines; two lateral spines on carapace. First pleopod of male with no shoulder on cephalic surface; tip terminating in four parts; mesial process spiculiform, directed caudodistad and slightly laterad; cephalic process hooding central projection over cephalodistal surface; caudal process small, arising from caudolateral surface at base of central projection; central projection conspicuous and directed caudodistad, being partially hooded by cephalic process. Annulus ventralis subovate, with greatest length in transverse axis; a long (usually multituberculate) finger-like projection from the sternum just cephalad of annulus, extending caudad on either side of midventral line, covers at least one-third of it; sinus forms an undulating line between the two finger-like projections and cuts the caudal margin of annulus.

Remarks.—My collection contains about 800 specimens from the panhandle of Florida, and I have seen a number of specimens from the extreme southern part of Alabama.

All of the specimens of *versutus* before me bear a weak median carina on the rostrum. In some specimens the rostrum is relatively short with a long acumen; in others it is much longer with a short acumen; in some the sides of the rostrum are almost parallel while in others they are more convergent. The antennal scale is sometimes narrow and without an angle on the mesiodistal margin; in other specimens it is broad and with a distinct angle. The areola is also very variable in its proportion of the length to width—the general tendency seems to be that in the western part of the panhandle it is broad and short, and becomes longer and narrower in the eastern part of the range, but this is not without exception. The form of the chelae in the western part of the range tends to be relatively long and slender with the fingers meeting along the entire length; specimens from the eastern part of the range, however, tend to have the palm broader and thicker, and the fingers gaping. In some specimens there is a single row of comparatively very long, recurved, sharp spines along the inner margin of the palm, while in others no single, well defined row is present, and in the occasional few which do have a single row, most of the tubercles are very low and rounded rather than spiny. Distinct variations occur in the development of spines on the carpus and merus of the first pereiopods. In some specimens the spines are low and blunt while in others they are very long, sharp, and slender. Either of the two rows of spines on the ventral side of the merus may be more strongly developed. Slight differences are seen in the curvature of the hooks on the ischiopodites of the fourth pair of pereiopods. The first pleopod of the male shows only slight variations; there is a shoulder on the lateral caudodistal margin in some specimens, but it is absent in others; the caudal process in most instances appears as a small corneous thumb-like projection, but in one or two specimens it is in the form of a plain triangle across the caudolateral face of the tip. The annulus ventralis, though showing many minor variations of form, maintains a rather constant pattern.

Both morphologically and ecologically *versutus* shows rather close affinities with *spiculifer*.

Specimens Examined.—I have examined a total of 971 specimens of *Procambarus versutus*, from Florida and Alabama; of these, 809 were collected in Florida. In ALABAMA the 162 specimens were taken from the following counties: Baldwin, Conecuh, Elmore, Escambia, Lee, and Mobile. In FLORIDA this species was collected from Escambia, Holmes, Liberty, Okaloosa, Santa Rosa, Walton, and Washington counties.

SEASONAL DATA

	Jan.	Feb.	Mar.	Apr.	May	June	July	Aug.	Sept.	Oct.	Nov.	Dec.
♂ I				44	1	35				11		20
♂ II				86	3	56				12		7
♀		1		127	7	67				25		21
♀ (eggs)				4		2						
♀ (young)				1								
♂ (immature)		5		152		26				12		31
♀ (immature)		11		150		21				21		34

Geographical and Ecological Distribution.—*P. versutus* is widespread over a large area in Alabama and in the panhandle of Florida, and has been reported from Louisiana (Penn 1941: 8) and Mississippi (Lyle 1938: 76). Like *spiculifer* this species is confined to streams, and its range is discussed in terms of the various drainage systems from which it has been collected. In Alabama it is known from tributaries of the Coosa, Tallapoosa (these join to form the Alabama River), Alabama, Conecuh, and Perdido rivers, with the northernmost records in Shelby and Lee counties. Although *versutus* has not been taken from the Choctawhatchee River in Alabama, it is known from some of its tributaries in Florida and very likely will be found to occur in its tributaries in southern Alabama.

The range of *versutus* in Florida is very discontinuous. It has been taken from the drainage systems of the Perdido, Escambia, Blackwater, Yellow, Choctawhatchee, and Apalachicola rivers. Perhaps this apparent discontinuity is due to collecting, but I believe that it is at least in part real and correlated with habitat requirements of the species. Since this form is very definitely a stream inhabitant and probably does not voluntarily leave flowing water it does not have access to many of the smaller streams east of the Choctawhatchee Bay. It is noteworthy that this species has been taken from Little Sweetwater Creek along the Apalachicola River, but it is apparently absent from many similar creeks along the river in this same area.

In Escambia County both *spiculifer* and *okaloosae* were taken with *versutus*. In Okaloosa County *spiculifer* was associated with it. In Santa Rosa County *diogenes* was taken from a burrow in a ditch which drained into the stream in which *versutus* was abundant. In Walton County *spiculifer* and *pycnogonopodus* were found inhabiting streams with *versutus*.

P. versutus is found in much the same type habitat as that occupied by *spiculifer*, i.e., a stream which has a moderate or swift current and in which the water is clear and cool. This type of stream usually has a rather high oxygen content, and since the areola of this species is so broad it is likely that the amount of gill space left would result in a demand for a habitat in which the oxygen content is relatively high. I have never taken *versutus* from a very large stream. A lack of an efficient method of collecting in these large streams may well be responsible for the sparsity of specimens from such localities. (*P. spiculifer, okaloosae, paeninsulanus,* and *fallax* are the only Florida species known to inhabit the larger rivers of the state.) Apparently once established in a stream *versutus* penetrates farther into the headwaters than does *spiculifer*. This species is abundant in the headwaters of Little Sweetwater Creek, Liberty County, while in the lower reaches of the same stream *spiculifer* seems to be the sole crayfish inhabitant. Little Sweetwater Creek is fairly typical of the streams which flow through the deep ravines on the east side of the Apalachicola. For the most part the water flows over a sandy bottom interrupted occasionally by a clay outcrop where the stream cuts more slowly and sometimes forms a small waterfall. Here and there above a waterfall a fallen limb has collected debris and so dammed a small pool. In these pools the crayfish congregate among the leaf debris. If one of these pools is approached quietly the chelae and the antennae of many crayfish may be seen extending from beneath the leaves, but any disturbance causes them to quickly disappear. A score of specimens may be taken from a single small pool. In some places the stream is only an inch or so deep while in others it reaches a depth of two or three feet; at night the crayfish are almost as abundant in one place as another, but during the day they are always at least partially hidden, and only rarely is one seen in open water. It may be added that in the headwaters of other small streams in the ravines along the Apalachicola River I found no *versutus* but instead a member of the genus *Cambarus, latimanus*.

P. versutus has been collected from streams in which there was little vegetation and from others in which vegetation completely covered the bottom. At one extreme is a creek just south of Flomaton, Alabama, where the vegetation is so luxuriant that in low water the channel is choked and braided by dense growth of plants. Here *versutus* was very common and could be easily taken by running a dip net through the vegetation along the sides of the numerous small channels. At the other extreme are the large, clear, swift creeks along the Choctawhatchee Bay. All are relatively shallow with an occasional deeper pool and generally with bare sand bottoms. Vegetation is almost confined to the banks which, except near the mouths, usually support a luxuriant grow of semi-aquatics. The mouths of these open creeks meander across the sandy beach in constantly shifting, shallow channels. The crayfish are found among the clumps of shore line vegetation, under logs, and in other debris scattered in the stream.

Several times I have found this species in the springs which fed small streams. The smaller specimens are particularly abundant in such situations. Occasionally *versutus* is found in slowly flowing streams, but the bottom is always sandy. Such streams are subject to flooding, and then the increased current probably washes out the collected detritis and keeps the creek bottom comparatively clean. In a side bend in a stream north of Olive, Escambia County, it was noted that *versutus* was confined to the swifter waters of the channel while *okaloosae* was the sole occupant of the still, vegetation-choked backwaters.

THE PICTUS GROUP

Diagnosis.—The first pleopod of the first form male terminates in four distinct parts (except in *seminolae* and *lunzi* where the caudal process is absent); mesial process long and spiculiform; cephalic process either long and spiculiform as in *seminolae* or shorter and spiniform, may be straight or slightly bent but always strongly developed; caudal process, a small corneous projection from the caudodistal surface, usually the most inconspicuous of the terminal processes; central projection large, corneous, sharp, and usually bent caudad distally. The areola is broad and short (except in *lunzi* and the cavernicolous forms); lateral spines on rostrum are well defined (except occasionally in *seminolae* and *lunzi*); hooks present on third and fourth pereiopods.

The *pictus* subgroup—*pictus, pubescens,* and *youngi*—inhabits streams in the Atlantic drainage from the southeastern part of South Carolina, southward to the St. Johns River in Florida and lowland streams flowing through the Apalachicola Flatwoods in the Gulf drainage. The *seminolae* subgroup—*seminolae* and *lunzi*—occurs in the flatwoods of southeastern Georgia and South Carolina and northeastern Florida, and the *lucifugus* subgroup—*lucifugus lucifugus, lucifugus alachua,* and *pallidus,* all cavernicolous—inhabits the underground water system of west-central Florida.

THE PICTUS SUBGROUP

Diagnosis.—The characters are those of the *pictus* group, excluding the albinistic cavernicoles and the *seminolae* subgroup in which the first pleopod terminates in three parts instead of four.

process non-corneous, spiniform, and directed caudodistad; caudal process present only as a rounded ridge across the caudodistal margin of the main shaft; central projection corneous, compressed laterally, and directed caudodistad. Annulus ventralis broadly spindle-shaped; sinus originates on midcephalic margin, curves gently caudodextrad, then sinistrad crossing midventral line, and finally caudodextrad cutting midcaudal margin of annulus.

Holotypic Male, Form I.—Body subcylindrical, slightly compressed laterally. Abdomen slightly narrower than thorax (.50-.55 cm. in widest parts respectively). Width and depth of carapace subequal in region of caudodorsal margin of cervical groove. Greatest width of carapace slightly caudad of caudodorsal margin of cervical groove.

Areola broad with six minute punctations in narrowest part. Cephalic section of carapace about 2.9 times as long as areola (length of areola 25.3% of entire length of carapace).

Rostrum excavate, reaching beyond distal segment of peduncle of antennule; acumen longer than base of rostrum; margins of rostrum subparallel, slightly convex distad of base; acumen set off by long lateral spines. Upper surface of rostrum with minute punctations. Marginal ridges well developed, sharp. Subrostral ridges weak and not evident in dorsal view.

Postorbital ridges well defined and terminating cephalad in strong spines. Suborbital angle obtuse, weak; branchiostegal spine strong. A single very strong lateral spine present on sides of carapace. Surface of carapace punctate dorsad and weakly granulate laterad.

Abdomen longer than thorax (1.7-1.46 cm.).

Cephalic section of telson with three spines in each caudolateral corner.

Epistome broadly triangular with a small cephalomedian projection.

Antennules of usual form. A strong spine present on ventral side of basal segment.

Antennae reaching to caudal margin of telson; antennal scale slender and long, broadest proximad of middle; cephalic margin of blade continuous with cephalomesial margin, sloping without any sharp curve along cephalomesial margin; spine on outer margin strong.

Chela long and slender, subcylindrical. Hand with minute squamous ciliated tubercles over entire surface. Inner margin of palm with a large number of minute denticles. No ridges present on finger. Fingers not gaping. Dactyl much shorter than inner margin of palm. Opposable margin of dactyl and immovable finger without tubercles but bearing crowded minute denticles. Other surfaces with setiferous punctations.

Carpus of first pereiopod about 1.7 times as long as broad; no longitudinal groove above, entirely punctate. One very long spine on upper mesiodistal margin and another on lower laterodistal margin.

Merus entirely punctate with a strong spine on upper distal surface, one or two on middle of lower surface and another on lower mesiodistal and laterodistal margins.

Hooks on ischiopodites of third and fourth pereiopods, those on fourth bituberculate. Bases of coxopodites of fourth pereiopods with strong and tuberculiform prominences. Those on fifth smaller and compressed.

First pleopod extending to base of second pereiopod when abdomen is flexed. Tip terminating in four parts. Mesial process non-corneous, blade-like, and directed caudodistad and somewhat laterad. Cephalic process non-corneous, spiniform, long, and directed caudodistad. Caudal process present only as a rounded ridge across caudodistal margin of the outer part of main shaft. Central projection corneous, compressed laterally, and directed caudodistad.

Male, Form II.—Differs from the first form male chiefly in the reduced size of the chelae and the marked reduction in size of the secondary sexual characteristics. The usual differences appear in the pleopods.

Allotypic Female.—Differs from the first form male in having a much slenderer and weaker chela, and in having only three spines in caudolateral corners of cephalic section of telson.

Annulus ventralis broadly spindle-shaped; sinus originates on midcephalic margin, curves gently caudodextrad, then sinistrad, crossing the midventral line, and finally caudodextrad cutting the midcaudal margin of the annulus.

Measurements.—Male (form I) Holotype: carapace, height .54, width .55, length 1.46 cm.; areola, width .11, length .37 cm.; rostrum, width .20, length .57 cm.; abdomen, length 1.70 cm.; right chela, length of inner margin of palm .52, width of palm .23, length of outer margin of hand 1.0, length of movable finger .41 cm. Female Allotype: carapace, height .56, width .56, length 1.48 cm.; areola, width .12, length .37 cm.; rostrum, width .20, length .57 cm.; abdomen, length 1.72 cm.; right chela, length of inner margin of palm .35, width of palm .15, length of outer margin of hand .54, length of movable finger .33 cm.

Type Locality.—Guard House Branch, about one mile west of Weewahitchka on State Highway 52, Gulf County, Florida. This is a small clear stream flowing through gently sloping flatwoods. Most of my specimens were taken from dead leaves and twigs in the quieter reaches of the stream.

Disposition of Types.—The male holotype, the female allotype, and a second form male paratype are deposited in the United States National Museum. Of the remaining paratypes one male (form I) and one female are deposited in the Museum of Comparative Zoology; one male (form I) and one female in the University of Michigan Museum of Zoology. One male (form I), one male (form II), and 35 females are in my personal collection at the University of Florida.

Remarks.—*Procambarus youngi* is known from only two localities in Gulf County, and there are no marked differences in the specimens from these two localities.

This species, although somewhat disjunct, has its closest affinities with *Procambarus pictus* and *Procambarus pubescens*, both of which occupy streams in the Atlantic drainage.

Specimens Examined.—I have examined a total of 44 specimens of this species, which were collected in April and November. First form males were found in April.

The localities from which *P. youngi* has been taken are as follows: Guard House Branch, one mile west of Weewahitchka on State Highway 52, and Wetappo Creek, 6.4 miles east of the Bay County line on State Highway 52.

Geographical and Ecological Distribution.—This species is known from the drainage systems of the Chipola River and Wetappo Creek, and I suspect that it is much more common in this region than is indicated by the locality records cited above. A thorough search of the streams of the Apalachicola flatwoods may show that it is much more widespread.

Guard House Branch, the type locality, is a small clear stream flowing through flatwoods bordering the Chipola River. The creek apparently originates in seepage areas and flows through a shallow valley in which *Magnolia virginiana, Cliftonia, Taxodium, Smilax,* and *Pinckneya* are common. The

Geographical and Ecological Distribution.—This species is known from only two underground bodies of water, but in one of them, Gum Cave, it has been abundant. Recently the roof of this cave has partially collapsed so that it is difficult to get down to the water. Also the *lucifugus* population has shown a marked decrease in the past three years. Gum Cave is a bat *(Myotis austroriparis)* infested cave some 130 feet deep, penetrating the Ocala limestone *(Eocene)*. At a depth of 75 feet from the cave entrance several large pools of apparently still water occupy the major portion of the cave bottom. The pool from which most of the crayfish were taken is about 50 by 7 feet and from 12 to 56 feet deep. The water is very clear and has a temperature of 70° F. The surface film is often thick with detritis, and the bottom of the pool has a layer of dark muddy silt which contains an abundance of insect parts. Roaches and beetles *(Dermestidae)* are common in the cave, but a large amount of the insect fragments are probably derived from the bat droppings, and the dead beetles and roaches which wash into the water provide a comparatively abundant food supply. The chicken snake, *Elaphe quadrivitata quadrivitata*, and the whip scorpion, *Thelyphonus giganteus*, have also been taken from the cave. Other animals living in and around the water are a blind amphipod *(Crangonyx hobbsi* Shoemaker), copepods, and springtails.

Troglocambarus maclanei is the only crayfish that has been collected with *P. lucifugus lucifugus*.

Procambarus lucifugus alachua (Hobbs)

Plate IX, Figs. 156-160; Map 9

Cambarus lucifugus alachua Hobbs 1940a, Proc. U. S. Nat. Mus. 89 (3097): 402-406, fig. 18.

Cambarus lucifugus alachua Hobbs 1940a: 387, 389, 398; Hobbs (in press-a).

Procambarus lucifugus alachua Hobbs (in press-c).

Diagnosis.—Albinistic, eyes reduced and with a small pigment spot; rostrum with lateral spines, broadest at base; areola narrow with one or two, if any, punctations in narrowest part; male with hooks on ischiopodites of third and fourth pereiopods; palm of chela of first form male never bearded within but with several irregular rows of tubercles; postorbital ridges terminating cephalad in spines; entire lateral portions of carapace studded with tubercles; a strong lateral spine present on either side. First pleopod of male, form I, reaching coxopodite of third pereiopod and terminating in four distinct parts directed caudodistad; mesial process almost thread-like while the other terminal elements are not markedly different from those found in *lucifugus lucifugus*. Annulus ventralis also essentially like that in *lucifugus lucifugus*.

Remarks.—The members of this species show marked variations. The large females of *P. lucifugus alachua* from Hog Sink have deeply excavate sternums which bear tubercles just anterior to the annulus ventralis, although this character is not so greatly developed as in *P. pallidus.*

The amount, distribution, and color of the eye pigment are decidedly variable. While the eyes of most of the specimens have a relatively small, concentrated pigment spot which is dark in color, there is a range from a diffuse, brownish-cream mass to a small concentrated black spot. (In several small females the spot is brilliant red.) In some specimens the pigment seems to be more deeply imbedded in the eye thus giving the appearance of being more diffuse and lighter in color. On others it seems to be at the surface, packed tightly into a small black body.

The rostra of *lucifugus alachua* are also highly variable in shape and in the presence of accessory lateral spines, i.e., in addition to the spines delimiting the acumen, others are often present along the margins.

The largest specimen of this species in my collection is a female from Hog Sink, the carapace of which measures 4.7 cm., and the carapace of the smallest first form male is only 2.1 cm.

In general the specimens from Goat Sink are much smaller than those from Hog Sink, and in several specimens the rostrum is almost like that of *lucifugus lucifugus* in that it is slightly broader in the middle than at the base.

The single female collected from Dudley's Cave is rather dark, creamy-brown (probably very old) with a relatively large, though lightly colored, pigmented area in the eye. The sternum is deeply excavate and tuberculate.

The apparent intergrades between *lucifugus lucifugus* and *lucifugus alachua* are known only from Indian Cave, Marion County, Florida (Hobbs 1940a: 406). The specimens here are also variable. Among them, however, are those which possess characters of both races, although in all of the specimens there is a small pigment spot in the eye. The first form male of this locality is unknown.

The relationships of this species are discussed under *P. lucifugus lucifugus* (p. 135).

Specimens Examined.—I have examined a total of 212 specimens of *Procambarus lucifugus alachua*, from the following localities in Alachua County: about 10 miles west of Gainesville (Sec. 24, Twp. 10S, R. 18E), Hog Sink; about 13 miles west of Gainesville (Sec. 32, Twp. 9S, R. 18E), Dudley Cave; about 11 miles northwest of Gainesville (Sec. 21, Twp. 9S, R. 18E), Squirrel Chimney; and about 12 miles northwest of Gainesville (Sec. 20, Twp. 9S, R. 18E), Goat Sink.

SEASONAL DATA

	Jan.	Feb.	Mar.	Apr.	May	June	July	Aug.	Sept.	Oct.	Nov.	Dec.
♂ I	5	2	4	3						2	2	1
♂ II	8	6	4	1		1				6	5	
♀	18	28	7	10						19	24	7
♂ (immature)	14			1							2	
♀ (immature)	20	6	4								5	

Geographical and Ecological Distribution.—Four sinks or caverns in Alachua County all within 10 miles of one another, are the only localities from which this subspecies has been taken. Although many other caves and sinks in this region have been visited, *lucifugus alachua* has not been found in them. Thus the known range of this subspecies seems to be within a small portion of the central-western part of Alachua County.

In Goat Sink *T. maclanei* was taken with *P. lucifugus alachua*; in Hog Sink *P. pallidus* was collected with it, and in Squirrel Chimney all three species occurred together.

All of the specimens of *lucifugus alachua* in collections have been taken from the water of solution cavities in the Ocala limestone. The pools which probably represent exposed areas of connected underground water systems are exposed in various sinks, caves, or combinations of sinks and caves that are briefly described below.

Hog Sink is a small sink hole at the bottom of which is a small cavern leading underground from the bottom of the sink proper. Here some 25 to 30 feet beneath the surface of the ground is a pool approximately 20 by 15 feet in area, and ranging from a shallow margin toward the entrance of the cavern to a depth of six or eight feet at the opposite side. The light in the cave is very dim, but with a flashlight one finds that the water is clear with an occasional thin film of detritis on the surface. The bottom consists of a layer of mud, sand, and silt, with an occasional limerock outcrop. Here the crayfish are always abundant, and with them has been taken a small fish, *Erimystax*, which is pigmented and apparently has functional eyes. On several occasions I have seen a very light-colored fish but have been unable to catch it. Annelid worms occur along the edge of the pool, and a few mosquitoes, terrestrial isopods, spiders, bats, and frogs have frequently been observed in the cave.

Like Hog Sink, Goat Sink is a combination sink hole and cave, though differing from the former in that the water extends out into the bottom of the sink proper. Thus it is exposed, at least in part, to direct daylight and shaded only by the high, steep walls and the trees overhanging the mouth of the sink. Near the edge of the pool the crayfish may be seen without the aid of a flashlight, but beneath the overhanging wall, where the water is deeper and extends back into the small cavern, a light is necessary. The pool is essentially like that of Hog Sink though its fauna is markedly different. Several white amphipods, *Crangonyx hobbsi* Shoemaker, and a few beetles are the only animals which have been observed in the water.

Dudley Cave is a small limestone cavern in which is located a single very small pool about four by five feet in size. This pool can be reached only by crawling and squeezing through a narrow passage sloping downward at about a 50 degree angle. The pool, which is in entire darkness, is very shallow and frequently becomes dry. Only two crayfish have been observed in this cave, and only one of these, a female, was captured. Along with this species an albino isopod, *Asellus hobbsi* Maloney, was taken.

Within 100 yards of this cave is another slightly larger one, containing a single deep pool of turbid water, about six feet in diameter. No crayfish have been observed in the pool, but scores of the isopod just mentioned have been seen hovering at the edge of the pool, and when a light was trained on them they moved rapidly into deeper water.

Intergradation between
Procambarus lucifugus lucifugus and *Procambarus lucifugus alachua*

From two caves in Marion County I have collected several crayfish that appear to be somewhat intermediate between these two subspecies. Since *P. lucifugus lucifugus* has been found in Citrus and Hernando counties and *P. lucifugus alachua* in Alachua County, it is logical that if there are transitional forms between these two, the Marion County region would be the place to expect them, and there is little doubt that these specimens from Waldo Cave (about 4 miles southwest of Ocala [State Highway 74]) and Indian Cave (about 7 miles southwest of Ocala [between State highways 16 and 74]) are intergrades *lucifugus* x *alachua*.

Procambarus pallidus (Hobbs)
Plate X, Figs. 161-165; Map 9

Cambarus pallidus Hobbs 1940a, Proc. U. S. Nat. Mus. 89 (3097): 394-398, fig. 16.
Cambarus acherontis Hobbs (not *acherontis* Lönnberg) 1937: 154.
Cambarus acherontis pallidus Hobbs 1938b: 90-91 (nomen nudum).
Cambarus pallidus Hobbs 1940a: 387, 389; Hobbs (in press-a).
Procambarus pallidus Hobbs (in press-c).

Diagnosis.—Albinistic, eyes reduced and without a pigment spot; rostrum with lateral spines, broadest at base; areola very narrow, almost obliterated in middle; male with hooks on ischiopodites of third and fourth pereiopods; palm of chela of first form male never bearded within but with two or three irregular rows of tubercles; postorbital ridges terminating cephalad in spines or tubercles; entire lateral portion of carapace studded with tubercles, a heavy bi- or trispinose tubercle replacing lateral spines. First pleopod of male, form I, reaching coxopodite of third pereiopod and terminating in four parts; cephalic process small, arising from cephalolateral margin of tip and directed caudomesiad, somewhat hooding the central projection; mesial process long and spiculiform, and directed caudad; caudal process forming a corneous curved blade along the caudomesial margin; central projection corneous, beak-like, and directed caudad. Annulus ventralis subovate with the greatest length in transverse axis; sinus arising on cephalic margin near midventral line, curves gently sinistrad, then somewhat sharply dextrad, crossing midventral line, finally caudodextrad, and terminating before reaching caudal margin of annulus near midventral line; sternum just cephalad of annulus with two multituberculate, finger-like projections extending caudad, in some specimens overhanging a portion of annulus.

Remarks.—Since the first form male is known from only two localities, Warrens Cave and Squirrel Chimney, variation is discussed only in connection with the females. The only marked difference in the females is in the development of the sternum just anterior to the annulus ventralis. In the specimens from Warrens Cave and Riverbed Cave the sternum is very deep and bears tubercles which extend caudomesiad and almost meet just anterior to the annulus ventralis. The sterna of the females from Hog Sink and Devils Hole bear tubercles which are not so well developed nor do they approach so closely along the midventral line.

P. pallidus has its closest affinities with the members of the *pictus* group, and it is probably most closely related to *P. lucifugus alachua*. The similarity in the development of the sternum to that of an undescribed species of this group from Toombs County, Georgia, is noteworthy.

Specimens Examined.—I have examined approximately 40 specimens, from seven localities in Alachua and Columbia counties, Florida. These localities are as follows: Columbia County—Riverbed Cave, about three miles west of High Springs [State Highway 5A]. Alachua County—Pallidus Sink (Sec. 15, Twp. 8S, R. 17E); High Springs Cave (Sec. 2, Twp. 8S, R. 17E); Squirrel Chimney (Sec. 21, Twp. 9S, R. 18E); Hog Sink (Sec. 24, Twp. 10S, R. 18E); Devils Hole (Sec. 18, Twp. 10S, R. 18E); Warrens Cave (Sec. 13, Twp. 9S, R. 18E).

SEASONAL DATA

	Jan.	Feb.	Mar.	Apr.	May	June	July	Aug.	Sept.	Oct.	Nov.	Dec.
♂ I			3	1						1		
♂ II	2		9	1							1	1
♀	2	1	4	1					1	2	2	
♂ (immature)												
♀ (immature)			2									

Geographical and Ecological Distribution.—This species has been taken from seven localities along the northeastern periphery of the lime sink region in Alachua and Columbia counties. *P. pallidus* has never been taken abundantly from any cavern, and one may visit any of these caves or sinks, except Squirrel Chimney, without seeing a single specimen. Seldom is it found in Warrens Cave and Hog Sink, and only rarely in Devils Hole. Several years ago numerous specimens were collected from the latter locality by Mr. O. C. Van Hyning, and since that time I have visited this locality several times, but only twice did I obtain specimens of *pallidus*.

This species is very rare in collections, although it is probably common in the underground waters in the northern part of Alachua and southern part of Columbia counties.

In Squirrel Chimney and Hog Sink *P. pallidus* was taken with *P. lucifugus alachua*, and in the former *T. maclanei* has been collected.

Warrens Cave, the type locality of this species, is one of the larger and better known caves in this section of the state. The opening to the cave is in the bottom of a sinkhole, and the passage downward is in the form of a fissure which extends upward within 15 to 20 feet of the surface. The single pool of water in the cave is about 60 to 75 feet below the ground level and is about nine by ten feet in size. Water appears in the cave intermittently, usually shortly after rains (after heavy rains it is evident that considerable water has drained into the cave from the outside). There is usually a suspension of silt in the water, and the bottom of the pool, which is littered with small limbs and sticks, is covered with a layer of silt. At one edge the pool is about a foot deep and drops off to a deep hole on the opposite side. The fauna of the cave is relatively poor; a few bats, frogs *(Rana sphenocephala)*, spiders, cave crickets, beetles, and moths are about all of the animals observed, and they occur in very small numbers.

The crayfish are usually seen resting on the bottom and move only when disturbed; then they dart around and attempt to retreat into the deep hole. The largest number of crayfish observed in the pool at one time was four; three of them were captured.

Riverbed Cave lies along the course of an old river bed which is now dry; the bare limestone walls occasionally reach a height of some 25 feet. Water has honeycombed the soft limestone, and numerous small caverns occur along the abandoned course. At one place a pit-like depression in the canyon floor exposes the edge of a pool which extends beneath an overhanging wall. Here the water is very clear and cool and is shaded from direct sunlight by the overhanging wall and a luxuriant growth of trees and ferns. At the edge the pool is very shallow but gradually deepens for some 25 feet, and then disappears into the honeycombed crevices. It is impossible to get beyond the very edge of the pool, for there is less than a foot of clearance between the water and the overhanging mass of rock. Several years ago a small opening above and a few feet to one side of this pool formed a passage which led down into a larger body of water continuous with the pool just mentioned. This larger body of water covered the entire floor of the fissure-like opening and was about three or four feet wide and 10 to 15 feet long. Since there was no ledge along the edge of this fissure-pool it was necessary to straddle it, and to capture a crayfish on the bottom, some 12 feet below, was very difficult. A recent dislodgment of the wall on one side of the fissure has made it impossible for an average-sized person to reach the water. Even when this fissure-pool was accessible I was able to catch only a few crayfish, for the passage was so small that in taking equipment down to the pool such a disturbance was set up that most of the crayfish had retreated into deeper parts before I had an opportunity to use what materials I could carry in. Other similar situations in the same riverbed apparently contain no crayfish. A white amphipod, *Cran-*

gonyx hobbsi Shoemaker, a few copepods, and several frogs, *Rana sphenocephala*, are the only forms I have seen in the pools.

Devils Hole differs from the other situations from which subterranean crayfish have been collected. It is in an open sink with an overhanging wall of limestone; the water is exposed to sunlight, and it is very dark in color as compared with the water in which all of the rest of my cavernicoles have been taken, and it is distinctly stagnant. Scattered over the muddy bottom of the pool which is shallow near the edge and progressively deeper under the overhanging wall are many rocks, dead tree trunks, and limbs. Among the animals inhabiting this pool are the same white amphipods, which were taken from the water when it was quite high, beetles (*Hydroporus clypealis* Shp.), mosquitoes, and frogs. On only one occasion have I collected *pallidus* in it, and then the water was much lower and clearer.

A description of Hog Sink is on page 138.

THE SEMINOLAE SUBGROUP

The characters are those of the *pictus* group except that the first pleopod of the first form male terminates in three rather than in four parts. No cavernicolous species are known.

This subgroup comprises two species, *P. seminolae* and *P. lunzi*. *P. seminolae* inhabits the flatwoods in the southeastern part of Georgia and the northeastern part of Florida. This species is by far the most ubiquitous member of the *pictus* group, inhabiting ponds, streams, and roadside excavations, and is apparently a capable burrower. *P. lunzi* (Hobbs 1940b) is more poorly known but has been found in a roadside ditch and in burrows near Early Branch, Hampton County, South Carolina.

Procambarus seminolae, sp. nov.

Plate X, Figs. 166-170; Plate XXIV; Map 9

Cambarus pubescens Hobbs (not of Faxon) 1937: 154.

Diagnosis.—Rostrum with or without small lateral spines; areola broad with three to five large punctations in narrowest part; male with hooks on ischiopodites of third and fourth pereiopods; palm of chela of first form male never bearded within but with an irregular row of eight or nine tubercles, usually subtended by other irregular rows; postorbital ridges terminating cephalad in spines or tubercles; a lateral spine may or may not be present on carapace. First pleopod of male, form I, reaching coxopodite of second pereiopod and terminating in three distinct parts; mesial process long and spiculiform and directed distad and caudolaterad; cephalic process also long and spiculiform and directed distad; central projection long, curved, blade-like, and directed caudodistad. Annulus ventralis subcylindrical or subovate with a submedian depression; sinus originates in the depression on or near midventral line, runs sinistrad, and makes a hairpin turn, crossing midventral line where it turns gently or abruptly back to midventral line, and proceeds along it to midcaudal margin of annulus.

Holotypic Male, Form I.—Body subovate, compressed laterally. Abdomen narrower than thorax (1.58-1.74 cm. in widest parts respectively). Width of carapace less than depth in region of caudodorsal margin of cervical groove. Greatest width of carapace about midway between cervical groove and caudal margin of carapace.

Areola broad with four punctations in narrowest part. Cephalic section of carapace 2.3 times as long as areola (length of areola 30.3% of entire length of carapace).

Rostrum almost flat above, reaching base of distal segment of peduncle of antennule; margins parallel for a short distance at base, but converging to base of acumen; the latter, however, not set off by lateral spines. Upper surface of rostrum with scattered punctations. Marginal ridges moderately strong. Subrostral ridges weak and not evident in dorsal view.

Postorbital ridge well defined and terminating cephalad in small tubercles. Suborbital angle acute; branchiostegal spine moderately strong. A large tubercle present on either side of carapace just caudad of cervical groove. Surface of carapace punctate dorsad, weakly granulate laterad.

Abdomen longer than thorax (3.90-3.70 cm.).

Cephalic section of telson with three spines in each caudolateral corner.

Epistome broadly subovate with a slight cephalomedian projection.

Antennules of usual form. A spine present on ventral side of basal segment.

Antennae extending caudad beyond caudal margin of telson; antennal scale broad, broadest in middle. Cephalomedian angle rounded; spine on outer margin moderately strong.

Chela long, heavy, subovate, slightly depressed. Hand with squamous ciliated tubercles over entire surface. Inner margin of palm with a row of nine or ten tubercles. A well defined submedian ridge present on upper surface of both fingers. Fingers not gaping. Opposable margin of dactyl with nine tubercles on proximal half, between and distad of which are crowded minute denticles. Lateral surface of dactyl with a row of five tubercles on basal half; distal half with setiferous punctations. Upper surface with a submedian ridge flanked proximad by tubercles and distad by setiferous punctations. Lower surface with tubercles on proximal half and setiferous punctations on both proximal and distal portions. Opposable margin of immovable finger with six tubercles on proximal half and a large one extending from lower margin at base of distal third. Crowded minute denticles between and distad of these tubercles. Lateral surface of immovable finger with a groove bearing setiferous punctations. Upper and lower surfaces similar to those of dactyl.

Carpus of first pereiopod about 1.6 times as long as broad; a well defined longitudinal groove above, punctate except on mesial and dorsomesial surfaces. Two strong tubercles on mesial surface and another on distal dorsomesial margin; two other prominent ones on ventrodistal and distal ventromesial margins.

Merus sparsely punctate on lateral and proximomesial surfaces, tuberculate otherwise. An irregular row of tubercles on upper surface with two distinctly larger ones on upper distal portion. Lower surface with an irregular row of about 15 tubercles and an inner row of about 18.

Simple hooks on ischiopodites of third and fourth pereiopods. Bases of coxopodites of fourth and fifth pereiopods with strong outgrowths. Those on fourth rounded and heavy; those on fifth compressed, rounded, and decidedly smaller.

First pleopod extending to base of second pereiopod when abdomen is flexed. Tip terminating in three parts. Mesial process corneous, spiculiform, and directed distad and caudolaterad. Cephalic process corneous, long, spiculiform, and directed distad. Central projection corneous, long, curved, blade-like, and directed caudodistad.

Male, Form II.—Differs from the first form male in having small lateral teeth on the rostrum; chela much smaller and slenderer; lateral spine on both sides of carapace; pleopod terminating in three parts, but none are corneous or spiculiform.

Allotypic Female.—Differs from the male, form I, in having proportionally shorter chela; lateral spines on side of carapace.

Annulus ventralis subcylindrical with a submedian depression with high lateral walls; sinus originates in the depression on midventral line, runs sinistrad, and makes a hairpin turn back to midventral line where it turns abruptly caudad to midcaudal margin of annulus.

Measurements.—Male (form I) Holotype: carapace, height 1.8, width 1.74, length 3.70 cm.; areola, width .26, length 1.12 cm.; rostrum, width .64, length 1.0 cm.; abdomen, length 3.9 cm.; right chela, length of inner margin of palm 1.39, width of palm 1.14, length of outer margin of hand 5.57, length of movable finger 1.96 cm. Female Allotype: carapace, height 1.90, width 1.76, length 3.89 cm.; areola, width .23, length 1.19 cm.; rostrum, width .64, length 1.08 cm.; abdomen, length 4.20 cm.; right chela, length of inner margin of palm .90, width of palm .84, length of outer margin of hand 2.40, length of movable finger 1.39 cm.

Type Locality.—Roadside excavation about nine miles northeast of Gainesville, on State Highway 13, Alachua County, Florida. This shallow excavation has standing water in it except during the very dry seasons and supports a few clumps of *Castalia*. Rotting logs and other debris are scattered over the muddy bottom.

Disposition of Types.—The male holotype, the female allotype, and a second form male paratype are deposited in the United States National Museum. Of the remaining paratypes one male (form I), one male (form II), and a female are deposited in the Museum of Comparative Zoology; one male (form I), one male (form II), and a female in the University of Michigan Museum of Zoology; one male (form I), one male (form II), and a female in the Philadelphia Academy of Natural Sciences. Ten males (form I), 47 males (form II), 76 females, 14 immature males, and 15 immature females are in my personal collection at the University of Florida.

Remarks.—Although the degree of variation in this species is marked, the first pleopod of the male remains decidedly constant. The annulus ventralis of the female, on the other hand, is highly variable. Not only do variations exist between specimens from two distant localities but often in a single small population. From a single locality there are specimens in which the rostrum is very broad and tapering and others in which it is narrow and with margins hardly converging. The rostra of some specimens are devoid of lateral spines; in others the spines are well developed. In some specimens the acumen is long and heavily pubescent, while in others it is either short and plain or not delimited by lateral spines. The ratio of the length of the cephalothorax to that of the areola ranges from 3.7 to 6.0; the antennal scale is distinctly variable, and the postorbital ridges terminate anteriorly either with or without spines. There are far too few specimens at hand to work out the significance of these many variations. Occasionally all of the specimens from a given locality show a marked uniformity and suggest that here one has encountered a definitely inbred population.

P. seminolae has its closest affinities with *P. lunzi*.

Specimens Examined.—I have examined a total of 552 specimens of *P. seminolae*, 265 of which were collected in Florida; the remaining 287 from the southeastern part of Georgia. The following constitute the state and county records for this species: GEORGIA—Appling, Ben Hill, Brooks, Camden, Clinch, Colquitt, Cook, Dooly, Echols, Glynn, Lowndes, and Wayne. FLORIDA—Alachua, Clay, Columbia, Duval, Hamilton, Nassau, and Union.

SEASONAL DATA

	Jan.	Feb.	Mar.	Apr.	May	June	July	Aug.	Sept.	Oct.	Nov.	Dec.
♂ I	1	2			1	12			2	10	8	1
♂ II		9	6		1	33		3		56	17	5
♀	2	15	5			44		2	6	68	41	12
♀ (eggs)										2	1	
♀ (young)												
♂ (immature)						5		2		73		2
♀ (immature)		1		2		7		2	2	76		6

Geographical and Ecological Distribution.—Since the above records are scattered over a comparatively wide expanse of territory in south-central and southeastern Georgia and in northeastern Florida, it is evident that they are too few to give more than an indication of the geographic and ecologic limits of the range. This is the more evident when it is considered that the southern limits, which are based upon more intensive and detailed collecting, do not coincide with any recognizable ecologic or geographic barrier and are totally ignored by *fallax* and *paeninsulanus*, the most common associates of *P. seminolae* in northern Florida.

In the Gainesville area, for example, abundant collecting has not produced any records for *P. seminolae* south of State Highway 41, though the ranges of *fallax* and *paeninsulanus* extend unbroken throughout the area.

P. seminolae has been collected with *P. paeninsulanus*, *fallax*, *pubischelae*, *advena*, *spiculifer*, and *pygmaeus*.

It is evident from the association of *seminolae* with the above species that it is nearly ubiquitous. It has been collected from clear, swift streams, stagnant pools, roadside ditches, flatwoods ponds, and burrows. Even the types of soils in which the burrowing specimens were taken were markedly different.

P. seminolae is most common in the flatwoods, where it inhabits both temporary ponds and ditches. During and after a rainy season it is abundant in the small flatwoods ponds. In a dry season numerous burrows are scattered over the bottom of the dried up ponds and crowded around the roots of trees or stumps which are found in the shallow basins. They often have an opening at the surface of the ground and another at the bottom of the ditch (the latter flooded after rains), but the burrow itself is always relatively simple. Although many of these burrows extend to a depth of two or three feet, it often happens particularly in dry seasons that the water table falls below this depth.

At such times *P. seminolae* is able to maintain itself in the moist bottom of the burrow for considerable periods. In the flatwoods ditches burrows are also numerous. The crayfish appear to inhabit the burrows even in seasons of abundant rainfall and leave them only at night. Even then they do not wander far from the mouth of the burrow and at the slightest disturbance scurry back to their holes.

In creeks *P. seminolae* is often abundant; especially is this true for the flatwoods streams of northeastern Florida where members of this species conceal themselves in vegetation or leaf drift. They are taken in small numbers from sand-bottom creeks and small rivers.

The only specimens that I have seen with eggs or young were taken from burrows, and it is not uncommon to find a female and a first form male together in a burrow.

Genus TROGLOCAMBARUS Hobbs

Genus *Troglocambarus* Hobbs (in press-c), Amer. Mid. Nat.

Diagnosis.—"First pleopod of first form male terminating in four distinct parts; the mesial and cephalic processes are somewhat spiniform and non-corneous; the caudal process and central projection are corneous and somewhat blade-like, the latter terminating in a rather sharp point. The eyes reduced and without pigment. The third maxilliped is tremendously enlarged, the ischiopodite of which bears no teeth along the inner [mesial] margin. The species is rather small and in life almost transparent" (Hobbs, in press-c).

This is a monotypic genus and further discussion will be found under the treatment of *Troglocambarus maclanei*.

Troglocambarus maclanei Hobbs
Plate I, Plate X, Figs. 171-175; Map 10

Troglocambarus maclanei Hobbs (in press-c), Amer. Mid. Nat.

Diagnosis.—Small, never exceeding two inches in length; carapace transparent; rostrum without lateral spines; areola of moderate width; hooks on ischiopodites of third and fourth pereiopods; palm of chela of first form male never bearded within, smooth; postorbital ridges terminating cephalad without spines or tubercles; no lateral spines present on carapace; third maxillipeds much enlarged and extending cephalad of peduncle of antennule and reaching distal end of carpus of chela when both are extended; mesial margin of ischiopodites bearing strong, long setae, but never teeth. First pleopod of first form male extending cephalad to coxopodite of fourth pereiopod and terminating in four distinct parts; mesial process long, spiniform, and directed caudolaterad; cephalic process spiniform, bent caudad, and extends somewhat mesiad of central projection; caudal process blade-like, corneous, and extends caudad from caudolateral tip of appendage; central projection corneous, acute, and directed caudodistad. Annulus ventralis subeliptical with the greatest length in the transverse axis, somewhat hidden beneath a flattened plate-like shelf arising from between the third and fourth pereiopods, plate deeply cleft; sinus originates in the cephalosinistral region, curves gently dextrad almost to midventral line where it turns caudosinistrad, then turns abruptly dextrad, crossing midventral line, and swings caudad and caudosinistrad to midcaudal margin of annulus.

Remarks.—*Troglocambarus maclanei* appears to have its closest affinities with the primitive *pictus* group of the genus *Procambarus*.

It was only after numerous visits to the localities from which this species is known that it was discovered. Because of its peculiar habit of clinging ventral-side-up to the submerged ceilings I had completely overlooked its presence in caves that I had visited scores of times, and it was only by accident that it was found at all. Mr. William M. McLane and I were collecting in Squirrel Chimney for *Procambarus pallidus* and had been disturbing the water with the dip net in attempting to catch several specimens. Having missed all of the large crayfish Mr. McLane scooped a small crayfish which he had taken for an immature into the dip net. In all probability this crayfish was frightened or torn away from its perch by currents and retreated to the bottom. As soon as it was realized that this was not a freak, another visit was made to the cave without finding a single specimen, and it was not until Mr. Lewis Marchand had dived below the submerged ceilings of this cavern and made a thorough search that it was discovered that *maclanei* is largely confined to the cavern ceilings that dip below the water level.

Specimens Examined.—I have examined a total of 28 specimens of *Troglocambarus maclanei*, which were collected in three localities, one of which is in Citrus County (Gum Cave, about six miles southwest of Floral City) and the other two in Alachua County (Squirrel Chimney [Sec. 21, Twp. 9S, R. 18E] and Goat Sink [Sec. 20, Twp. 9S, R. 18E]).

Specimens have been collected only during the months of March, May, and September, and first form males were found in March.

Geographical and Ecological Distribution.—While this species is known from only three localities it has the largest range of any of the Florida cavernicolous species. It is probable that *maclanei* is much more common than my collections indicate, but because of its apparently peculiar habitat and habits, additional specimens and localities will be slow to accumulate. However, it is reasonable to suppose that this species must be confined to the subterranean waters of the western part of the peninsula and does not extend much beyond the present known range of the cavernicolous species of the *pictus* group.

Troglocambarus maclanei has been collected with *P. lucifugus lucifugus*, *P. lucifugus alachua*, and *P. pallidus*.

"Squirrel Chimney," the type locality of this species, is a circular solution cavity with almost vertical walls, the latter supporting a luxuriant growth of liverworts, mosses, and small ferns. This chimney penetrates the surface

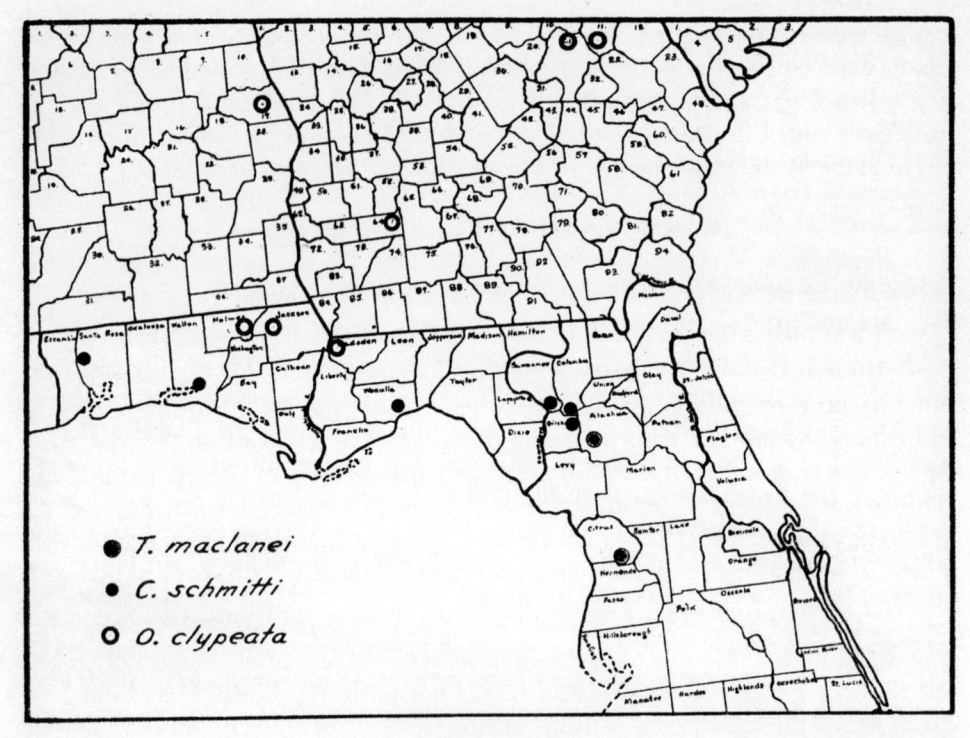

Map 10.—Distribution of the Genera *Troglocambarus*, *Cambarellus*, and *Orconectes* in Florida.

soil and limestone to a depth of approximately 50 feet, where it strikes subterranean water. Debris has fallen into the sink and has accumulated at the water level so that a little less than half of the bottom area is open water, the rest of it being covered with fallen leaves which are supported by dead tree trunks and limbs. Within six to eight feet of the bottom a small opening about three feet in diameter leads out into a fissure about 25 yards long and four feet wide, the whole bottom of which is filled with water ranging in depth from a few inches to 30 feet at the deepest place sounded. The light is very dim inside of the fissure, and a short distance inside of it, it is completely dark. The water is very clear, but the surface film sometimes supports a coat of fine silt and debris. The bottom consists of mud, sand, and silt, with large limerock outcrops.

"The crayfish are always found in the shallower portions of the pool, where they cling bottom-side-up [ventral-side-up] to the limestone roof where it dips below the water table. Of the 24 specimens taken only two were found on the bottom, the rest clinging to the submerged ceiling. Two or three specimens were taken from open water as they released their hold on the ceiling or were jarred loose, and floated toward the bottom. To collect most of the specimens it was necessary for Mr. Marchand to dive below the surface and explore the submerged ceilings, and in some instances to go horizontally beneath these ceilings as much as 15 feet. To facilitate such a method of collecting a headlight and water-goggles were used. Mr. Marchand found that *T. maclanei* could be taken readily by hand, whereas it is almost impossible to catch the larger cave species without a dip net" (Hobbs, in press-c).

GENUS CAMBARELLUS ORTMANN

Subgenus *Cambarellus* Ortmann 1905a, Proc. Amer. Phil. Soc. 44 (180): 97.
Genus *Cambarellus* Hobbs (in press-c), Amer. Mid. Nat.

Diagnosis.—"First pleopod of first form male terminating in three distinct parts; the cephalic process is always absent; the three terminals may be spiniform, somewhat truncate, spatulate, or even trough-like. In the male hooks are present on the ischiopodites of the second and third pereiopods. This genus includes a group of very small crayfish; specimens seldom reach a length of two inches. Third maxillipeds proportionally of normal size, with a row of teeth along the inner margin of the ischiopodite" (Hobbs, in press-c).

"*Cambarellus* has its main abode in Mexico, and only one species is known from a single locality in Louisiana (New Orleans)" (Ortmann 1906: 20). "The most primitive species of the subgenus *Cambarellus (C. shufeldti)* is found in Louisiana. *C. chapalanus* appears slightly more primitive compared with *C. montezumae* and its varieties, and is found in western Mexico. Of the *montezumae* forms, *areolatus* is the most primitive and the most northern, nearest to the United States, while *occidentalis* is the most advanced (shape of rostrum), and is western in Mexico. Thus the evidence is partly contradictory. Leaving out *chapalanus,* the general trend of the evidence is to show that the subgenus originated in the southern United States and immigrated into Mexico, first into the central plateau, then into the Pacific slope" (ibid.: 24).

A new species of *Cambarellus, schmitti,* has been found to occur in Florida, thus extending the range of the subgenus some 500 miles east of New Orleans.

Cambarellus schmitti[15], sp. nov.

Plate X, Figs. 176-180; Plate XIX; Map 10

Diagnosis.—Small, never exceeding one and one-half inches in length; rostrum with lateral spines; areola broad with four or five punctations in narrowest part; male with hooks on ischiopodites of second and third pereiopods; palm of chela of first form male never bearded within, smooth; postorbital ridges terminating cephalad in long spines which extend cephalad of

---

[15]This species is named in honor of Dr. Waldo L. Schmitt of the United States National Museum, whose aid to my work on Florida crayfishes has been invaluable.

the cephalolateral margin of carapace; a strong lateral spine on either side of carapace; first pleopod extending to coxopodite of third pereiopod and terminating in three distinct parts; mesial process non-corneous, thumb-like, swollen, and directed caudodistad; caudal process and central projection corneous, slender, acute, and bent caudodistad; mesial process does not form a longitudinal trough as found in the *montezumae* subspecies and *chapalanus*. Annulus ventralis movable, compressed cephalocaudad, rounded ventrad with angular lateral margins; sinus originates in a tubercle on cephaloventral surface sinistrad of midventral line, curves caudad across the ventral surface where it turns sharply dextrad to midventral line; sternum immediately caudad of annulus raised into a large acute tubercle which fits into a shallow groove when annulus is bent caudad.

Holotypic Male, Form I.—Body in cross section subovate, slightly compressed laterally. Width of abdomen and thorax subequal. Width of carapace in region of caudodorsal margin of cervical groove slightly greater than depth. Greatest width of carapace slightly caudad of caudodorsal margin of cervical groove.

Areola wide; three times as long as wide; 28.9% of entire length of carapace; punctate (about five in narrowest part). Sides subparallel in middle.

Rostrum broad, of moderate length; surface flat and with setiferous punctations; sides subparallel; ridges not prominent; lateral spines long and slender. Acumen subspiculiform, extending cephalad almost to base of flagellum of antennule.

Postorbital ridge terminates cephalad in long spines which extend cephalad beyond cephalolateral margin of carapace. Substrostral ridge visible for some distance cephalad in dorsal aspect. Surface of carapace punctate; lateral portions weakly granulate. Lateral spines present. Suborbital angle acute and prominent; branchiostegal spine strong.

Abdomen longer than thorax.

Cephalic section of telson with two spines in each of the caudolateral corners.

Epistome broadly subtriangular, flattened.

Antennules of usual form. A strong spine present on mesioventral margin of basal segment.

Antennae extending caudad to caudal margin of second abdominal segment; antennal scale large, broadest in middle; spine on outer margin strong, extending to tip of rostrum.

Chela subcylindrical, long, slender, with setiferous punctations; not tuberculate. Inner margins of both fingers with minute denticles and hairs. No ridges on either finger.

Carpus longer than broad with scattered setiferous punctations. A sharp spine present on lower distal edge, and also one on dorsomesial distal margin.

Merus with one strongly developed acute spine on ventral surface slightly distad of midlength. Upper surface with a strong spine on distal third; a very strong spine on distal lateral edge.

Hooks on ischiopodites of second and third pereiopods; hooks strong and bituberculate. Proximal tubercles of hook on second pereiopod acute, distal one more inconspicuous; both tubercles of hooks on third pereiopods acute and strong. Coxopodite of fourth pereiopod with a prominent hook directed caudoventrad.

First pleopod reaching to coxopodite of third pereiopod and terminating in three distinct parts. Mesial process non-corneous, large, heavy, and tapering. Cephalic process lacking. Caudal process arises from caudolateral margin and is subspiculiform and corneous. Central projection corneous, slender, and shaped somewhat like a scythe blade. Mesial and caudal processes, while straight, are directed caudad at about 45 and 30 degree angles to the main shaft respectively.

Male, Form II.—Essentially like male, form I; first pleopod non-corneous and with processes reduced; all spiny portions are also decidedly less conspicuous.

Allotypic Female.—Width of abdomen greater than thorax.

Annulus ventralis broad with margin rounded; sternum immediately caudad of annulus raised into a large tubercle which fits into a shallow groove when annulus is bent caudad; sinus originates on cephalic face of annulus, sinistrad of midventral line, curves caudad and dextrad across median ridge, then dextrad to about midventral line on caudal face.

Measurements.—Male (form I) Holotype: carapace, height .34, width .38, length .83 cm.; areola, width .08, length .24 cm.; rostrum, width .14, length .24 cm.; abdomen, length 1.10 cm.; right chela of paratype, length of inner margin of palm .27, width of palm .16, length of outer margin of hand .55, length of movable finger .24 cm. Female Allotype: carapace, height .52, width .50, length 1.09 cm.; areola, width .08, length .34 cm.; rostrum, width .15, length .30 cm.; abdomen, length 1.26 cm.; right chela, length of inner margin of palm .29, width of palm .20, length of outer margin of hand .64, length of movable finger .29 cm.

Type Locality.—Small spring flowing into the Suwannee River at Branford, Suwannee County, Florida. Most of my specimens were taken from dense growths of *Cabomba* and *Riccia*.

Disposition of Types.—All of the type series were collected from the type locality and from Blue Springs, Gilchrist County, Florida. The male holotype, the female allotype, and a second form male paratype are deposited in the United States National Museum. Of the remaining paratypes one male (form I), one male (form II), and a female are deposited in the Museum of Comparative Zoology; a similar series in the University of Michigan Museum of Zoology and the Charleston Museum. Twenty-one males (form I), 16 males (form II), 25 females, and two immature males are in my personal collection at the University of Florida.

Remarks.—The rostra of the Walton County specimens are slightly longer and narrower than in specimens from the more eastern localities, but on the whole my specimens show very little differences.

C. schmitti has its closest affinities with *shufeldtii* from Louisiana and *montezumae* from Mexico. The first pleopod of the male more nearly resembles *montezumae*, for the terminal processes are slightly recurved, not straight as in *shufeldtii*. There are strong lateral spines on the carapace which are also present in *shufeldtii* but absent in some *montezumae*.

Ortmann put forth the theory (labeling it as such) that, due to the primitiveness of the morphological characters and the peculiar discontinuity of the distribution, the probable origin of the subgenus *Cambarellus* was in the southern part of the United States, that it migrated into Mexico and there gave rise to the Mexican species of the subgenus. The discovery of *C. schmitti* neither refutes this theory nor substantiates it.

Specimens Examined.—I have a total of 198 specimens in my collection, all except 13 of which were collected in Florida. These 13 specimens were taken in Mobile County, Alabama. In Florida *schmitti* is found in the following localities: several springs on the Santa Fe River, in the southern part of Columbia County, and the northern part of Gilchrist County; Branford, Suto the subgenus *Faxonella, Orconectes (Faxonella) clypeata* (Hay). The other subgenus, *Orconectes,* extends into the southeast as far as northwestern Georgia, northern Alabama, and Mississippi.

Subgenus FAXONELLA Creaser

Subgenus *Faxonella* Creaser 1933b, Occ. Pap. Mus. Zool., Univ. Mich. (275): 21.

"In view of the peculiarities of the sexual appendages with one long ramus and one short one, I deem this species worthy of subgeneric ranking and designate it *Faxonella,* new subgenus of *Faxonius*" (Creaser 1933b: 21). *(Faxonius* is a synonym of *Orconectes.)*

Orconectes (Faxonella) clypeata (Hay)

Plate XI, Figs. 181-185; Map 10

Cambarus clypeatus Hay 1899, Proc. U. S. Nat. Mus. 22 (1187):122-123, fig. 2, nos. 1-4.

Cambarus clypeatus Ortmann 1902: 277, 279; Harris 1903: 58, 83, 137, 143, 153; Faxon 1914: 410, 426; Creaser 1933a: 17, 29, 36, 40, figs. 13, 15, 28.

Faxonius (Faxonella) clypeatus Creaser 1933b: 19, 20, 21, Pl. I, figs. 7, 8; Pl. II, figs. 1, 2.

Orconectes (Faxonella) clypeata Hobbs (in press-c).

Diagnosis.—Rostrum without lateral spines, broadest distad of base; areola wide with about five punctations in narrowest part; male with hooks on ischiopodites on third pereiopods only; palm of chela of first form male never bearded within but bearing a large number of small tubercles; postorbital ridges terminating cephalad without spines; no lateral spines on sides of carapace. First pleopod of first form male reaching coxopodite of first pereiopod and terminating in two rami; mesial process very short, acute, and noncorneous; central projection very long, slender, and corneous. Annulus ventralis subovate with the greatest length in transverse axis (see Plate XI, figure 183 for sinus pattern).

Remarks.—Although the Florida specimens show considerable minute variations these are slight when compared with the range in variation found in series of many other Florida species. In the first pleopod of the male the Florida specimens show a definitely smaller quotient (1.23) in the length of the basal portion divided by the length of the central projection than do Louisiana specimens (2.65); Alabama specimens (1.36) are intermediate, and northeast Georgia specimens have a quotient of 1.26. The trend toward a longer central projection toward the southeastern part of the range is definite. The annulus ventralis also exhibits a decided variation, but no definite limit to the variations has been marked for any given locality. Other variations may be observed in the rostrum.

The relationships of *clypeata* are given by Creaser (1933b: 21), as follows: "In the original description Hay suggested that the species, when the male was discovered, would probably prove to be a form related to *Cambarellus cubensis*. Faxon (1914: 401) felt inclined to group it with *Cambarus (Bartonius)*. This crayfish can now be correctly assigned to the genus *Faxonius*, but it is certainly an orphan among this group of crayfishes."

Specimens Examined.—I have examined a total of 176 specimens of *Orconectes clypeata* from Florida all collected during April; among them are 36 first form males.

The following localities are known from Florida: Gadsden County—about 7.5 miles west of Quincy on U. S. Highway 90. Holmes County—small pond near Bonifay; Holmes Creek between Bonifay and Chipley on U. S. Highway 90. Jackson County—about 2.5 miles south of Graceville, State Highway 52; about 4.5 miles west of Sneads, U. S. Highway 90; 2.4 miles southeast of Florida-Alabama line on U. S. Highway 231.

Geographical and Ecological Distribution.—*Orconectes clypeata* has been recorded from the following states and counties: ALABAMA—Mobile County (Creaser 1933b: 21); LOUISIANA—Grant Parish (Creaser 1933b: 21); MISSISSIPPI—Hancock County (Hay 1899: 122); OKLAHOMA—La Flore County (Creaser 1933b: 21). From my collection the following may be added: ALABAMA—Butler and Lee counties; ARKANSAS—Greene and Philips counties; GEORGIA—Dougherty, Emanuel, and Jenkins counties.

The range of this species in Florida is again probably larger than is apparent from the known localities, but I believe that it is relatively safe to assume that the eastern boundary extends no farther than the tributaries of the Ochlocknee River in this state. On the basis of the occurrence of *O. clypeata* in the stream bed of Holmes Creek and other records near Graceville, Jackson County, it seems very probable that it occurs in the southeastern part of Alabama. I have never taken this species south of U. S. Highway 90, but judging from the range of *P. hubbelli*, a species found in the same habitat with *clypeata* in Holmes County, and the ranges of other consocies in Gadsden and Jackson counties, I believe that the range of this species extends southward to the coastal flatwoods. A small pond near Bonifay is the westernmost locality known for Florida. Since, however, it has been collected in Mobile County, Alabama, it is possible that additional intervening localities have been overlooked.

In Gadsden County this species was collected with *spiculifer* and *leonensis*; in Jackson County it was collected with *paeninsulanus*, and in Holmes County with *hubbelli*.

My specimens have been collected in at least four types of situations. In Jackson County they were taken from a roadside ditch in a small area of flat-with the *bartonii* or *diogenes* section than with the section of *hamulatus*; therefore I propose that the *bartonii* section be redefined to exclude "Eyes well developed."

This section now includes three of the Florida species: *C. latimanus, C. floridanus,* and *C. cryptodytes.*

The *bartonii* section occupies a large area in eastern North America extending southward from Canada throughout the Appalachian and Piedmont regions into the panhandle of Florida, and westward beyond the Mississippi River.

The Florida representatives of this section occupy portions of the drainage systems of the Apalachicola and the Ochlocknee rivers and part of the subterranean water system in Jackson County.

C. latimanus and *floridanus* are either primary or secondary burrowers while *cryptodytes* is a true cavernicole.

Cambarus latimanus (LeConte)

Plate XI, Figs. 186-190; Map 11

Astacus latimanus LeConte 1856, Proc. Acad. Nat. Sci. Philad. 7: 402.
Astacus latimanus Hagen 1870: 9, 10, 79.
Cambarus latimanus Hagen 1870: 78, 80, 82, 83-84, 98, 100, 105, 106, 107, Pl. I, figs. 43-46; Pl. III, fig. 162; Faxon 1884: 144; Faxon 1885a: 63, 64, 65, 69-70, 169, 173, 174, 178, 179, Pl. II, fig. 3; Faxon 1885b: 359; Underwood 1886: 370; Faxon 1898: 650; Hay 1902b: 436; Ortmann 1902: 277; Harris 1903: 58, 106, 107, 138, 142, 143, 144, 146, 147, 151, 152, 153, 154, 156; Ortmann 1905a: 119, 120, 122, 128, 129, 130, 135; Faxon 1914: 394, 395, 396, 425; Newcombe 1929: 279; Ortmann 1931: 106, 107, 108, 124-125, 141, 142, 143; Fleming 1939: 299, 300, 301, 302, 311, 314, 319, 320, Pl. XVIII, figs. 1-4; Carr 1940: 4; Hobbs 1941b: 110, 113, 118.

Diagnosis.—Rostrum without lateral spines; areola moderately broad with three or four punctations in narrowest part and length 28-33% of entire length of carapace; male with hooks on ischiopodites of third pereiopods only; chela compressed with two rows of five or six tubercles along inner margin of palm; postorbital ridges terminating cephalad with or without tubercles; several tubercles present on sides of carapace just caudad of cervical groove. First pleopod of male, form I, reaching coxopodite of third pereiopod and terminating in two distinct parts which are bent at right angles to the main shaft; mesial process bulbiform and non-corneous; central projection scythe-like and corneous. Annulus ventralis subovate with the greatest length in the transverse axis and with a submedian longitudinal depression; sinus originates near midlength and slightly dextrad of midventral line, curves gently caudodextrad, then somewhat sharply sinistrad, bending caudosinistrad just before crossing midventral line, after which it turns caudodextrad to cut midcaudal margin of annulus.

Remarks.—I am unable to distinguish specimens taken from the ravine at Camp Torreya from specimens from the type locality. I have about 60 specimens from the type locality, Athens, Clarke County, Georgia, and the only variations between these and Camp Torreya, Florida, specimens are slight differences in the rostra and areolae; specimens from either locality are as different among themselves as they are from each other.

From a ravine in the northern part of Torreya State Park where *latimanus* occurs I have several specimens which were taken from complex burrows and which differ from *latimanus* or *floridanus* in several respects. These specimens, instead of being gray with lighter gray or greenish markings or dull red, were a distinct orange-red. Anatomically, they differ slightly from *latimanus*. The length of the areola of one mature female is 38.5% of the entire length of the carapace whereas in *latimanus* it is only 28-33% (in this character it approaches *floridanus*); the rostrum is broader and shorter; the cephalic region of the carapace is proportionately higher than the thoracic region; the suborbital angle is round so that its presence is hardly discernible whereas in *latimanus* it is small but angular. This form seems to be distinct, but until further specimens have been procured it seems to be advisable to do nothing more than to call attention to its presence in the state.

C. latimanus has no close relatives among the Florida crayfish with the possible exception of *C. floridanus*. Its closest relatives are probably other members of the *bartonii* complex in the southern Appalachian region.

Specimens Examined.—I have a total of 133 specimens, of which 41 were collected in Liberty and Jackson counties, Florida, and the others from the following counties in Georgia: Clarke, Floyd, Green, McDuffie, Oconee, and Washington.

SEASONAL DATA

	Jan.	Feb.	Mar.	Apr.	May	June	July	Aug.	Sept.	Oct.	Nov.	Dec.
♂ I	3			3					2		1	
♂ II	8			6	2	1		9	9	2	2	2
♀	8			7	4	7		8	10		3	
♂ (immature)	18					1					1	2
♀ (immature)	11								3		1	

Geographical and Ecological Distribution.—In addition to the localities cited above the following have been published: ALABAMA—Etowah County (Faxon 1898); Lee County (Faxon 1914); FLORIDA—Liberty County (Carr 1940); GEORGIA—Baldwin County (Hagen 1870); Clarke County (Faxon 1885a); Cobb County (Faxon 1885a); MISSISSIPPI—(?)Ocean Springs (Faxon 1885a); NORTH CAROLINA—Wake County (Ortmann 1931); SOUTH CAROLINA—(Hagen 1870); Greenville County (Faxon 1885a); Richland County (Faxon 1885a).

C. latimanus has a relatively large range, and present records indicate discontinuity. Probably as is the case for many of the other seemingly discontinuous ranges, a lack of intensive collecting is responsible.

The range of *latimanus* in Florida seems to be largely restricted to the ravine streams along the Apalachicola River. In Jackson County this species was collected from a small stream near the river. The ravines in Liberty County are cut into the high bluffs on the east side of the river and extend northward from near Bristol, Florida, into southwestern Georgia. Collections have been made in several of these ravines, but *latimanus* has been found in only two of them.

" . . . 'Camp Torreya' is a deep ravine situated about one mile south of Rock Bluff Landing—one of many cutting back into the sandy uplands which border the Apalachicola River valley on the east. Moist and cool, these ravines contain an interesting assemblage of plants and animals which comprise southern species, northern relicts, and a number of endemics of very limited distribution. The steep slopes and bottoms of the ravines are covered with dense forest in which beech, magnolia, red oak, hickories and sweet gums are dominant, and ironwood, wild plum, slippery elm, holly, spruce pine and other trees are also common. The undergrowth is composed largely of stinking cedar or savron *(Tumion taxifolium,* formerly *Torreya)*, a coniferous shrub or small tree confined to the Apalachicola ravines, as is also the less common Florida yew *(Taxus floridana)*. A characteristic member of the ground flora is the needle palm *(Rhapidophyllum hystrix)*, growing side by side with such northern types as bloodroot, Trillium, hepatica and toothwort. The sandy soil is thickly carpeted with dead leaves and humus, and brush heaps and decaying logs are numerous. The ravines are noted for their botanical peculiarities; Thone (1927) has given a popular account of them based largely upon and illustrated by excellent photographs of the one at Camp Torreya; Harper (1914) has published a more detailed treatment with lists of the plants" (Hubbell 1939: 159).

The rill and brook occupying this ravine flow into Sweet Water Creek, which a mile or so below empties into the Apalachicola River. In the ravine brook piles of debris, rocks, and limbs have accumulated, and among these the crayfish are most abundant. Also along this stream are many complex burrows which extend downward and beneath its banks. Because there is such an entwined mass of roots in the soil, digging into these burrows is a laborious job that seldom yields any specimens.

In June 1938 I was collecting from this stream, and while digging around among the roots in the creek bed my attention was attracted by a scraping noise higher up on the bank, and upon investigating I found that it was a large female *latimanus* walking about at the mouth of a burrow that opened three feet above the level of the stream.

C. latimanus seems to be fairly abundant in this stream, and I have seen many burrows along the banks of similar streams in other ravines but have succeeded in capturing only four specimens in one of the other streams.[16] There are no *latimanus* in the headwaters of Little Sweetwater, but I saw many burrows which were possibly constructed by this species along Kelly Branch, a small tributary of this stream.

The only species taken from the ravine brooks with *C. latimanus* was *C. diogenes*.

Cambarus floridanus Hobbs

Plate XI, Figs. 191-195; Map 11

Cambarus floridanus Hobbs 1941b, Amer. Mid. Nat. 26 (1): 114-118, Pl. I, figs. 1, 4, 5, 8, 9; Pl. II, figs. 16, 19, 22, 25, 31, 32.

Cambarus floridanus Hobbs (in press-c).

Diagnosis.—Rostrum without lateral spines; areola narrow with one or two punctations in narrowest part and length 37.2-41.1% of entire length of carapace; male with hooks on ischiopodites of third pereiopods only; chela compressed with two rows of five to seven tubercles along inner margin of palm; postorbital ridges terminating cephalad with or without small spines or tubercles; no lateral spines present, instead a single small tubercle on either side of carapace just caudad of cervical groove. First pleopod of male, form I, reaching coxopodite of third pereiopod and terminating in two distinct parts which are bent at right angles to the main shaft; mesial process bulbiform and non-corneous except at tip; central projection scythe-like and corneous. Annulus ventralis roughly subrectangular; sinus originates cephalad along midventral line, turns sharply dextrad, then caudad for a short distance, then sharply sinistrad, passing the midventral line where it turns sharply caudad to cut midcaudal margin of annulus.

Remarks.—There are considerable variations in this species, but I am unable to recognize any peculiar differences between the specimens from any of the four localities from which it has been taken.

This species is clearly a member of the *bartonii* group which is common in the Appalachian region. Perhaps its closest affinities are with *C. latimanus*; this, however, is questionable.

Specimens Examined.—I have examined a total of 39 specimens of *Cambarus floridanus*, from the following counties in Florida: Gadsden, Jackson, and Leon. These specimens were collected during the months of March, May, November, and December, and first form males were collected during the latter two months.

Geographical and Ecological Distribution.—*C. floridanus* is known from four localities in the Apalachicola region. Three of these are in the drainage

---

[16] This stream is in a deep ravine in the northern part of Torreya State Park.

THE DIOGENES SECTION

Ortmann (1905a: 119) defined the *diogenes* section as follows: "Carapace ovate, compressed, without lateral spines. Rostrum without marginal spines. Chelae short and broad, depressed, ovate. Areola very narrow or obliterated in the middle, always distinctly longer than half of the anterior section of the carapace. Eyes well developed."

This section includes three of the Florida species: *diogenes, byersi,* and another which I am not naming because the first form male is still unknown.

The *diogenes* section is widely distributed throughout the United States east of the Mississippi River and is represented in Arkansas, Colorado, Iowa, Kansas, Minnesota, Missouri, Nebraska, and Oklahoma.

In Florida representatives of this section occur in scattered localities from the Perdido River eastward to the ravines along the east bank of the Apalachicola River.

C. diogenes is a secondary burrower; *byersi* is a primary burrower; and the unnamed species, although poorly known, is apparently a secondary burrower.

Cambarus diogenes diogenes (Girard)

Plate XII, Figs. 201-205; Map 11

Cambarus diogenes Girard 1852, Proc. Acad. Nat. Sci., Philad. 6: 88.
Cambarus diogenes Girard 1852: 88, 89, 90; Hagen 1870: 6, 7, 9, 75, 82, 83; Smith 1874: 639; Forbes 1876: 5, 19; Bundy 1877: 171; Bundy 1882: 183; Faxon 1884: 115, 117, 144, 145; Abbott 1884: 1157-1158; Faxon 1885a: 18, 28, 48, 56, 58, 63, 69, 70, 71-75, 76, 77, 78, 92, 94, 100, 160, 173, 174, 176; Faxon 1885b: 359; Faxon 1890: 624, 625; Hay 1891: 147; Hay 1896: 487; Shufeldt 1896: 85-89; Faxon 1898: 649, 650; Osborn and Williamson 1898: 21; Hay 1898a: 939, 961; Harris 1900: 267; Ortmann 1902: 277, 283; Hay 1902c: 235; Harris 1903: 59, 61, 62, 63, 72, 76, 81, 83-85, 96, 101, 137, 138, 140, 141, 142, 143, 144, 145, 146, 147, 148, 149, 150, 151, 152, 153, 154, 155, 157, 158, 163, 164, 165, 167, 168, 169, Pl. IV, fig. 7; Ortmann 1905a: 95, 117, 119, 120, 122, 123, 124, 125, 126, 128, 129, 135-136; Ortmann 1906b: 349, 351, 376, 386, 391, 393, 402-410, 414, 415, 416-424, 452, 453, 456, 457, 457-462, 463, 464, 465, 466, 476, 480-486, 488, 489, 490, 492, 496, 497, 500, 501, 503, 509-512, 521, Pl. A, fig. 3; Pl. XXXIX, fig. 11, Pl. XL, figs. 6, 7; Pl. XLI, figs. 5, 6, 7, 8; Ortmann 1905d: 387, 388, 389, 394, 398-400, 404, 405; Ortmann 1907: 706, 707, 709, 711, 712, 713, 714, 715, 716, 2 figs.; Williamson 1907: 749, 750, 751, 755, 758, 759-763; Pearse 1910a: 70; Pearse 1910b: 10, 11, 12, 15, 20, Pl. 8; Fowler 1911: 341, 348-352; Cockrell 1912: 51; Chidester 1912: 288, 289; Faxon 1914: 394, 395, 400, 426;

Hay and Shore 1918: 400, 401, Pl. 28, fig. 4; Hay (in Evermann) 1918: 232, 233, 234; Engle 1926: 88, 89, 92, 93, 94, 95, 96, 97, 98, 99, 100, 101, figs. 1, 4, 5; Newcombe 1929: 278, 279, 280, 286, 287, fig. 1; Creaser 1931c: 260, 261, 263, 267, 269, 270, fig. 36, map 7; Creaser 1931b: 9; Ortmann 1931: 95, 105, 145, 147, 149, 151, 152-155, 156, 157; Creaser 1933a: 16, 18, 40, fig. 2; Fleming 1939: 300, 301, 302, 303, 311, 317, 318, 319, Pl. XXII, figs. 1, 4; Creaser 1931a: 243; Creaser 1932: 321, 324, 331, 335, 336, figs. 5, 11; Hobbs 1941b: 110, 121; Hobbs and Marchand (in press).

(?) *Cambarus nebrascensis* Girard 1852: 91; Hagen 1870: 8, 83, 85, 98, 102, 107; Faxon 1884: 145; Faxon 1885a: 75-76, 160, 174; Harris 1903: 109, 151, 153; Faxon 1914: 426; Engle 1926: 88.

Cambarus obesus Hagen 1870: 43, 53, 70, 77, 80, 81-83, 88, 98, 99, 100, 101, 102, 105, 106, 107, 108, 109, Pl. I, figs. 39-42, Pl. III, fig. 163, Pl. IX; Abbott 1873: 83; Faxon 1884: 144; Faxon 1885a: 58, 71, 92; Harris 1903: 80, 88, 93, 94, 167; Faxon 1914: 426; Engle 1926: 92.

Diagnosis.—Rostrum without lateral spines, long and tapering; areola obliterated in middle; male with hooks on ischiopodites of third pereiopods only; chela heavy with one or two rows of squamous tubercles along inner margin of palm, outer finger distinctly flattened at base; postorbital ridges terminating cephalad with or without small tubercles; no lateral spines present, instead a few tubercles present on either side of carapace just caudad of cervical groove. First pleopod of male, form I, reaching coxopodite of third pereiopod and terminating in two distinct parts which are bent at right angles to the main shaft; mesial process, while slender and for the most part corneous, occasionally terminates in one or more small corneous teeth; central projection blade-like. Annulus ventralis subrectangular with the diagonal in the longitudinal axis; sinus originates on midventral line slightly caudad of cephalic margin of annulus, curves caudodextrad, then caudad and caudosinistrad to midventral line where it turns sharply caudad and terminates just before reaching midcaudal margin of annulus; fossa disappears below dextral wall in middle of annulus.

Remarks.—It seems likely that the present systematic status accorded to *Cambarus diogenes* will prove unsatisfactory when a comparative study of specimens from the entire range has been made, and that it will be necessary to recognize several geographic races.

In the Florida specimens there are variations in the rostrum, chela, and annulus ventralis, and the length of the areola ranges from 36.9 to 42.4% of the entire length of carapace. The same variation in color pattern is noted in the Florida specimens that is described by Hobbs and Marchand (in press) for *diogenes* in the Reelfoot Lake area.

Ortmann (1905a: 123) points out that *C. diogenes* is closely related to *C. uhleri, fodiens* (of which *argillicola* is a synonym), *carolinus*, and *monongalensis*.

Specimens Examined.—I have examined approximately 40 specimens of *Cambarus diogenes* from the following counties in Florida: Calhoun, Escambia, Holmes, Jackson, Liberty, Okaloosa, Santa Rosa, and Washington. These specimens were collected during the months of April, May, October, November, and December, and first form males were taken only during December. One female with eggs was found in April, and another female with young in May.

Geographical and Ecological Distribution.—Besides the above county records for Florida *C. diogenes* is known to occur in the following states: Alabama, Arkansas, Colorado, Delaware, District of Columbia, Georgia, Illinois, Indiana, Iowa, Kansas, Kentucky, Maryland, Michigan, Minnesota, Mississippi, Missouri, Nebraska, New Jersey, North Carolina, Ohio, Oklahoma, Pennsylvania, Tennessee, Virginia, West Virginia, Wisconsin, and Wyoming.

The range of *diogenes* in Florida is confined to the panhandle west of the Ochlocknee River, where it seems to be restricted to the drainage systems of the major streams. It has been taken from the Apalachicola, Choctawhatchee, Shoal, Escambia, and Perdido river systems, but I do not know the drainage systems to which one of the Santa Rosa County localities belongs, nor that for one of the Washington County localities.

The habits of this species have been observed by several authors, and Harris (1903: 85-86) gives an excellent summary of observations up to that time.

P. hubbelli was taken from burrows adjacent to those of *diogenes* in Washington County, and both *hubbelli* and *paeninsulanus* were taken from burrows near those of *diogenes* in the same locality. *P. versutus, paeninsulanus, hubbelli, spiculifer,* and *pycnogonopodus* have been taken from streams and ditches adjacent to where *diogenes* was collected.

In the several localities from which *diogenes* was taken in Florida the burrows were above the low water level in stream beds or ditches. In most instances the burrows have only one or two branches from the vertical passageway, and a slender, high chimney is built at the mouth of the passage. The chimney appears to be carefully constructed of small, round mud, or clay pellets. When the mouths of burrows have more than one opening there may or may not be a chimney over each of them.

Along the banks of a small stream in Early County, Georgia, each burrow had one opening below the water level, penetrated for about a foot into the bank, and usually continued to an above water opening by way of a single passage upward. Over some of these were small chimneys. When the crayfish were dug out it was found that they were usually below the water table.

In the flood plain of Holmes Creek at Graceville chimneys of hundreds of burrows of *diogenes* were scattered over the flats. The burrows were only about two and one-half feet deep and extended about six inches below the water table. In Washington County burrows of *diogenes* were found in an open seepage area along a drainage ditch. The entire slope was wet and soft, and the soil, though mucky at the surface, had a subsoil of clay and sand.

At least some of the deep ravines along the east bank of the Apalachicola (p. 160) are inhabited by *diogenes*. Here the crayfish dig their burrows along the steep banks of the small stream. Since this species has been taken from these situations it seems probable that the numerous burrows in the flood plain of the river in Liberty County were probably those of *diogenes*. Here the burrows are very deep, and some of them are four inches in diameter.

In the Escambia River drainage this species is abundant. Here in the flood plains of the river itself and its small tributaries chimneys constructed by *diogenes* are numerous. Their chimneys range in height from a few inches to more than a foot; many are built at the bases of trees or cypress "knees" and are partially supported by them. In one or two instances specimens were taken outside of the burrow in swamp pools; one of these specimens had just moulted.

Cambarus byersi Hobbs
Plate XII, Figs. 206-210; Map 11

Cambarus byersi Hobbs 1941b, Amer. Mid. Nat. 26 (1): 118-121, Pl. I, figs. 6, 10, 14; Pl. II, figs. 17, 20, 23, 26, 27.
Cambarus byersi Hobbs (in press-c).

Diagnosis.—Rostrum without lateral spines, short and broad; areola obliterated in middle; male with hooks on ischiopodites of third pereiopods only; chela strongly depressed with a single cristiform row of eight or nine tubercles along inner margin of palm; mesial margin of dactyl of first pereiopod with a distinct excision slightly proximad of midlength; postorbital ridges terminating cephalad without spines or tubercles; no lateral spines or tubercles present. First pleopod of male, form I, reaching coxopodite of third pereiopod and terminating in two distinct parts which are bent at more than a right angle to the main shaft; mesial process twisted, grooved, non-corneous, and not bulbiform; central projection blade-like and corneous. Annulus ventralis subquadrate with the longitudinal axis passing through opposite angles; cephalic margin fused with sternum of thorax just cephalad of annulus; median surface flattened with the highest ridges on the caudolateral sides, the highest caudodextrad; cephalolateral ridges diminish cephalomesiad so that the cephalic angle is almost as flat as the median portion (only a low mound present at cephalic angle); caudodextral wall overhangs the fossa which disappears slightly sinistrad of the midventral line near caudal tip of annulus; sinus evident only in caudosinistral region where it cuts across the caudosinistral ridge.

Remarks.—The specimens in my collection show relatively little variation. Slight differences are seen in the shape of the rostrum and in the chela, but no local variants are known to exist.

The relationship of this species is definitely with *C. fodiens*, and perhaps subsequent collecting between Florida and Louisiana may disclose the presence of intergrades between these two species. *C. fodiens* is known to occur from Michigan southward into Louisiana, and it does not seem unlikely that this species is an offshoot of the *fodiens* stock.

Specimens Examined.—I have examined a total of 64 specimens of *Cambarus byersi* of which 20 were collected in the following counties in Florida: Escambia, Okaloosa, and Santa Rosa. The remaining 44 specimens were taken in Baldwin and Escambia counties, Alabama. Specimens have been collected only during the months of April, May, and June, and first form males were taken in April and June.

Geographical and Ecological Distribution.—This species appears to be confined to the drainage systems of the Perdido, Escambia, Blackwater, and Yellow rivers where it constructs complex burrows in meadows and seepage areas along the small tributaries. It is almost certainly a primary burrower. The burrows are marked by well formed chimneys, and there may be as many as four or five chimneys opening into the same complex system of passageways. Some of the tunnels extend horizontally for four or five feet from a central deep passage, and many of them dip beneath the water table several times. Like most burrowing crayfish this species appears to seek out tangles of roots among which it digs its intricate tunnel system. A 10% return when digging for this species is about as good as can be expected.

Cambarus (Species incertis)

Plate XII, Figs. 211-214

The crayfish here referred to undoubtedly constitute a new species, but because I do not have a mature male it seems best to do nothing more than briefly describe the female, and give the exact localities from which the species has been taken, along with a few observations I have made concerning its habits.

One of my immature males is far enough along in its development that the species can with certainty be referred to the genus *Cambarus*.

Diagnosis of Female.—Rostrum without lateral spines, margins subparallel or slightly converging; areola obliterated in middle; chela depressed, inner margin of palm with two rows of six or seven tubercles, movable finger with a deep excision near midlength, immovable finger heavily bearded at base; postorbital ridges terminating cephalad with or without tubercles; no lateral spines on carapace, one or two larger tubercles present instead; ischiopodites of third maxillipeds densely bearded. Annulus ventralis similar to that found in *diogenes* (Pl. XII, fig. 213).

Specimens Examined.—I have examined a total of 19 specimens of this species, all of which were collected in the Escambia River flood plains in Escambia and Santa Rosa counties.

The localities from which this species has been taken are as follows: along the Escambia River on State Highway 62, west of Jay, Santa Rosa County; McCaskill's Mill Creek, about 12 miles southwest of Jay, Santa Rosa County; and in a slough of the Escambia River near Bluff Springs, Escambia County.

Ecological Observations.—This species was not discovered until April 1942, and at this time the Escambia River was flooded and spread out over a great part of its flood plain. My first specimens were taken along with *P. shermani* and *P. blandingii acutus* in a seine which was dragged through small swamp pools along McCaskill's Mill Creek. In collecting at night a number of these crayfish were seen at the mouths of burrows which were scattered over the floor of the pools. Failing to get a first form male on this trip, a second attempt was made in May after the flood water had largely subsided; however, this time I failed to get a single mature specimen. A diligent search was made in several localities, and, while a fair series of immature specimens were taken, not one mature individual was seen. A number of burrows were dug, but all of these turned out to have been constructed by *Cambarus diogenes*. I have no explanation to offer as to why the mature specimens were not in evidence at this time.

Wakulla: *P. kilbyi, P. paeninsulanus, P. rogersi campestris, P. spiculifer, C. schmitti.*

Walton: *P. apalachicolae, P. hubbelli, P. pycnogonopodus, P. spiculifer, P. versutus, C. schmitti.*

Washington: *P. hubbelli, P. paeninsulanus, P. pycnogonopodus, P. spiculifer, P. versutus, C. diogenes.*

LITERATURE CITED

Abbott, C. C. 1873. Notes on Habits of Certain Crayfish. *Amer. Nat.*, Salem, Mass., 7: 80-84.

    1884. Are the "Chimneys" of Burrowing Crayfish Designed? *Amer. Nat.*, 18: 1157-1158.

Bundy, W. F. (Forbes, S. A.) 1876. List of Illinois Crustacea with Descriptions of New Species. *Bull. Ill. Mus. Nat. Hist.*, 1: 3-25, 1 pl.

    1877. On the Cambari of Northern Indiana. *Proc. Acad. Nat. Sci., Philad.*, 29: 171-174.

    1882. A List of the Crustacea of Wisconsin, with Notes on Some New or Little Known Species. *Trans. Wis. Acad. Sci., Arts and Letters*, 5: 177-184.

Carr, A. F. 1940. A Contribution to the Herpetology of Florida. *Univ. Fla. Pub., Biol. Series*, 3 (1): 1-118.

Chidester, F. E. 1912. The Biology of the Crayfish. *Amer. Nat.*, 46: 279-293.

Cockerell, T. D. A. 1912. The Fauna of Boulder County, Colorado. II. *Univ. Colorado Studies*, 9 (2-3): 41-52, 1 pl.

Cooke, C. W. 1925. The Coastal Plain in The Physical Geography of Georgia. *Bull. Geol. Surv. Ga.*, 42: 19-54.

    1939. Scenery of Florida. *Bull. Fla. St. Geol. Surv.*, 17: 1-118, 58 pls.

Cooke, C. W. and Mossom, S. 1929. Geology of Florida. *20th Ann. Rep. Fla. St. Geol. Surv.*, pp. 31-227, 29 pls.

Cope, E. D. 1872. On the Wyandotte Cave and its Fauna. *Amer. Nat.*, 6: 406-422, figs. 109-116.

Creaser, E. P. 1931a. Some Coinhabitants of Burrowing Crayfish. *Ecology*, 12 (1): 243-244.

    1931b. Three New Crayfishes from Puebla and Missouri. *Occ. Pap. Mus. Zool., Univ. of Mich.*, (224): 1-10, pls. 1-5.

    1931c. The Michigan Decapod Crustaceans. *Pap. Mich. Acad. Sci., Arts and Letters*, 13: 257-276, figs. 31-40, maps 3-8.

    1932. The Decapod Crustaceans of Wisconsin. *Wis. Acad. Sci., Arts and Letters*, 27: 321-338, figs. 1-13, 1 table.

    1933a. (and Ortenburger, A. I.). The Decapod Crustaceans of Oklahoma. *Univ. Okla. Pub., Biol. Surv.*, 5 (2): 14-47, figs. 1-19, 23-33.

    1933b. Descriptions of Some New and Poorly Known Species of North American Crayfishes. *Occ. Pap. Mus. Zool., Univ. Mich.*, (275): 1-21, 2 pls.

1934. A New Crayfish from North Carolina. *Occ. Pap. Mus. Zool., Univ. Mich.*, (285): 1-4, 1 fig.

Engle, E. T. 1926. Crayfishes of the Genus *Cambarus* in Nebraska and Eastern Colorado. *Bull. Bur. Fish.*, 42: 87-104.

Erichson, W. F. 1846. Ubersicht der Arten der Gattung Astacus. *Arch. Für Naturgeschichte, Zwölfter Jahrgang Erster Band mit Zwölf Knpfertaflen,* 12 (pt. 1): 86-103.

Faxon, W. 1884. Descriptions on New Species of *Cambarus*; to which is Added a Synonymical List of the Known Species of *Cambarus* and *Astacus*. *Proc. Amer. Acad. Arts and Sci.*, 20: 107-158.

1885a. A revision of the Astacidae. *Mem. Mus. Comp. Zool., Harvard Coll.*, 10 (4): 1-186, 10 pls.

1885b. A List of The Astacidae in the United States National Museum. *Proc. U. S. Nat. Mus.*, 8 (23): 356-361.

(in Garman) 1889. Cave Animals in Southwestern Missouri. *Bull. Mus. Comp. Zool.*, 12 (6): 237, pl. 1, figs. 1, 2, 3, 7; pl. II, fig. 1.

1890. Notes on North American Crayfishes, Family Astacidae. *Proc. U. S. Nat. Mus.*, 12 (785): 619-634.

1898. Observations on the Astacidae in the United States National Museum and in the Museum of Comparative Zoology, with Descriptions of New Species. *Proc. U. S. Nat. Mus.*, 20 (1136): 643-694, pls. 62-70.

1914. Notes on the Crayfishes in the United States National Museum and the Museum of Comparative Zoölogy with Descriptions of New Species and Subspecies to which is Appended a Catalogue of the Known Species and Subspecies. *Mem. Mus. Comp. Zool., Harvard Coll.*, 40 (8): 347-427, pls. 1-11.

Fleming, R. S. 1939. The larger Crustacea of the Nashville Region. *Journ. Tenn. Acad. Sci.*, 13 (4): 296-314, Oct., 1938; 14 (2): 261-264, Apr., 1939; 14 (3): 299-324, July, 1939.

Forbes, S. A. 1876. List of Illinois Crustacea with Descriptions of New Species. *Bull. Ill. St. Lab. Nat. Hist., Urbana*, 1 (1): 3-25.

Fowler, H. W. 1911. Crustacea of New Jersey. *Rep. N. J. St. Mus.* 1911, Part 2: 31-650, pls. 1-150.

Gibbes, L. R. 1850. On the Carcinological Collections of the United States, and an Enumeration of the Species Contained in them, with Notes on the Most Remarkable and Descriptions of New Species. *Proc. Amer. Ass.*, 3: 167.

Girard, C. 1852. A Revision of the North American Astaci, with Observations on Their Habits and Geographical Distribution. *Proc. Acad. Nat. Sci., Philad.*, 6: 87-91.

Hagen, H. A. 1870. Monograph of the North American Astacidae. *Ill. Cat. Mus. Comp. Zool., Harvard Coll.*, (3): 1-109, pls. 1-11.

Harper, R. M. 1914. Geography and Vegetation of Northern Florida. *Fla. Geol. Surv., 6th Ann. Rept.* pp. 163-451, figs. 40-90.

1925. Generalized Soil Map of Florida. *Fla. Geol. Surv.*

1906a. Mexican, Central American, and Cuban Cambari. *Proc. Wash. Acad. Sci.*, 8: 1-24, figs. 1-4.

1906b. The Crawfishes of the State of Pennsylvania. *Mem. Carnegie Mus.*, 2 (10): 343-521, 7 pls.

1907. Grabende Krebse in Nordamerika. Aus der *Natur-Sonderabdruck* aus Jahrg., pp. 706-716, figs. 1-5.

1931. Crawfishes of the Southern Appalachians and the Cumberland Plateau. *Ann. Carnegie Mus.*, 20 (2): 61-160.

Osborn, R. C. and Williamson, E. B. 1898. The Crayfish of Ohio. *6th Ann. Rep. Ohio Acad. Sci.*, p. 21.

Packard, A. S. and Cope, E. D. 1881. The Fauna of the Nickajack Cave. *Amer. Nat.* 15: 872-882.

Pearse, A. S. 1910a. A Preliminary List of the Crustacea of Michigan. *Rep. Mich. Acad. Sci.*, 41: 68-76.

1910b. The Crawfishes of Michigan. *Mich. St. Biol. Surv.*, Pub. 1, pp. 9-22, pls. 1-8.

Penn, G. H. 1941. Preliminary Report of a Survey of the Crawfishes of Louisiana. Abst. of Pap. 88th Ann. Meet. N. O. Acad. Sci. p. 8.

Rhoades, R. 1941. Notes on Some Crayfishes from Alabama Caves, with the Description of a New Species and a New Subspecies, *Proc. U. S. Nat. Mus.*, 91 (3129): 141-148, figs. 35, 36.

Rogers, J. S. 1933. The Ecological Distribution of the Craneflies of Northern Florida. *Ecol. Monog.*, 3 (1): 1-74, 25 figs.

Shufeldt, R. W. 1896. The Chimneys of Burrowing Crayfish. *The Observer*, 7: 35-89.

Smith, S. I. 1874. The Crustacea of the Freshwaters of the Northern United States, *Rep. U. S. Fish Comm.* (1872-1873), pp. 637-665.

Steele, M. 1902. The Crayfishes of Missouri. *Pub. Univ. Cincinnati*, Bull. 10: 1-50, 10 pls.

St. John, E. P. 1936. Rare Ferns of Central Florida. *Amer. Fern Journ.*, 26 (2): 41-55.

Thone, F. 1927. A Stranded Company. *Amer. For. and For. Life*, 33: 532-534, illus.

Underwood, L. M. 1886. List of the Described Species of Freshwater Crustacea from America North of Mexico. *Bull. Ill. St. Lab.*, 2: 323-386.

Ward, H. B. and Whipple, G. C. 1918. Freshwater Biology. John Wiley and Sons, Inc. N. Y., pp. i-ix, 1-1111, figs. 1-1534.

Williamson, E. B. 1907. Notes on the Crayfish of Wells County, Indiana, with Description of a New Species. *31st Ann. Rep. Dept. Geol. and Nat. Res. of Ind.*, pp. 749-763, 1 pl.

INDEX

Florida genera in bold-face capitals; new species in bold-face lower-case; synonymous names and obsolete combinations, italicized; diagnoses, descriptions, and discussions of species, genera, etc., bold-face numerals; occurrence in key, italicized numerals; plates and figures designated by Roman numerals followed by Arabic numerals.

Acknowledgments, 1.
Advances into Subterranean Waters, 12.
advena, Astacus, 75.
Alabama, counties collected from, 4.
Aquatic Habitats, see Ecology.

Barriers to Migration, 14.
Bartonius, 22, 155, 157.
blandingii, Astacus, 157.
Burrowing Habits: primary, secondary, tertiary, 20.

CAMBARELLUS, 149, 11, 22, 23, *28*.
   Cambarellus, 33, 155.
   Cambarellus, subgenus, 11, 22, 149, 151, 157.
   chapalanus, 149, 150.
   cubensis, 155.
   montezumae, subspecies, 150, **151**.
   schmitti, 149-153, *28*, 4, 15, 21, 95, 113, 118, 123, 148, 170, 171, 172; X: 176-180, XX.
   shufeldtii, 151.
Cambarinae, **23**, 1, 10, 12.
CAMBARUS, 156-157, *28*, 12, 22, 23, 155, 163, 168.
   Cambarus, subgenus, 33, 157.
   acherontis, 91, 134, 139.
   acherontis pallidus, 139.
   acuminatus, 157.
   acutissimus, 94.
   acutus, 94.
   advena, 75.
   advena geodytes, 80.
   affinis, 8, 9.
   alleni, 8, 9, 69.
   argillicola, 165.
   ayersii, 163.
   barbatus, 8, 9, 39, 46
   bartonii, 8, 135, 157.
   Bartonii Group, 161, 163.
   Bartonii Section, **157-158**, *32*.
   blandingii, 157.
   blandingii acutus, 94.
   byersi, **167-168**, *32*, 9, 13, 14, 20, 49, 103, 109, 156, 157, 164, 170, 171; XII: 206-210.
   cahni, 163.
   carolinus, 75.
   clarkii paeninsulanus, 9, 104.
   clypeatus, 154.
   cryptodytes, **162-163**, *32*, 9, 12, 21, 156, 157, 158, 171; XI: 196-200.
   diogenes diogenes, **164-167**, *32*, 15, 20, 64, 78, 95, 128, 156, 157, 161, 162, 169, 170, 171, 172; XII: 201-205.
   Diogenes Section, **164**, *32*, 158.
   evermanni, 8, 9, 107.
   fallax, 8, 9, 111.
   floridanus, **161-162**, *32*, 9, 14, 20, 156,

CAMBARUS, (cont.)
   157, 158, 159, 170, 171; XI: 191-195.
   fodiens, 165, 168.
   Fodiens Stock, 168.
   Group I, Faxon, 33, 153.
   Group II, Faxon, 33.
   Group III, Faxon, 33, 153.
   Group IV, Faxon, 153.
   Group VI, Faxon, 153.
   Group I, Hagen, 33, 153.
   Group III, Hagen, 33, 153.
   hamulatus, 163.
   Hamulatus Section, 158.
   hubbelli, 9, 67.
   kilbyi, 9, 64.
   latimanus, **158-161**, *32*, 14, 20, 21, 128, 156, 157, 171; IX: 186-190.
   lecontei, 8, 9.
   lucifugus alachua, 9, 136.
   lucifugus lucifugus, 9, 134.
   monogalensis, 165.
   montezumae, 149.
   nebrascensis, 165.
   obesus, 165.
   pallidus, 9, 139.
   penicillatus, 39, 40.
   pictus, 9, 130.
   pubescens, 142.
   rathbunae, 9, 59.
   rogersi, 9, 89.
   setosus, 163.
   shufeldtii, 149.
   species incertis, **168-169**, *32*, 12, 14, 20, 64, 95, 103, 156, 170, 171; XII: 211-214.
   spiculifer, 8, 119.
   stygius, 94.
   uhleri, 165.
   versutus, 8, 9, 126.
Caudal Process, 27.
Centrocaudal Process, 27.
Centrocephalic Process, 27.
Cephalic Process, 27.
clypeatus, Faxonius, 154.
Collecting, see Methods of
Counties collected from, 2.

Derivation of Fauna, 10; Cambarellus, 11; Cambarus 11, 12; Orconectes, 11; Procambarus, 11; Troglocambarus, 11.

Ecology, 16; Flatwoods and Seepage Areas, 19; Lenitic Situations, 18; Lotic Situations, 16; Subterranean Situations, 19, 136, 138, 139, 141, 142, 147, 148.
Ecological Habits and Habitats, 20; Burrows, 20; Outside of Burrows, 21; Subterranean Waters, 21.

INDEX

Endemism, 12.

Faxonella, **154**, *28*.
Faxonius, genus, 22, 153, 154, 155.
Faxonius, subgenus, 22, 153, 157.
Field Notes, 6, 7.
Flatwoods, see Ecology.
Florida, counties collected from, 2.
Florida as a Faunal Region, 9; Geological History, 9, 10; Physiography, 10.
Form I, 25.
Form II, 25.

Generic and Subgeneric Changes, Summary, 22.
Georgia, counties collected from, 4.
Girardiella, 22.

Key, 28.
Key Characters, 25; Carapace, 27; Pereiopod, 27; Pleopod, 25, 26; Rostrum, 27; Secondary sexual, 27.

latimanus, Astacus, 158.
Lenitic Situations, see Ecology.
Literature Cited, 172.
Lotic Situations, see Ecology.

Male, Form I, Form II, 25.
Material Studied, 3.
Measurements, 24.
Mesial Process, 27.
Methods of Collecting in: Open Water, 4, 5; Burrows, 5; Caves, 4, 7; Traps, 7.
Migration Routes, 13.

ORCONECTES, 153-154, *28*, 11, 22, 23.
 clypeata, **154-156**, *28*, 14, 15, 20, 21, 106, 148, 170, 171; XI: 181-185.
  Faxonella, subgenus, **154**, *28*.
Ortmannicus, 22, 33, 157.

Paracambarus, genus, 22, 23.
Paracambarus, subgenus, 22.
penicillatus, Astacus, 39.
penicillatus, Palinurus, 40.
Pereiopod, 25, 27.
Pleopod, First, 25; text figure, 26.
Preservation and Study, 7.
Previous Work, 8.
Principal Migration Routes, 13.
PROCAMBARUS, 33, *28*, 11, 22, 23, 35, 147.
 acherontis, **91-92**, *28*, 8, 9, 12, 13, 21, 83, 120, 121, 171; VI: 86-90.
 Acherontis Section, **91**, *28*.
 advena, **75-80**, *30*, 13, 15, 20, 23, 45, 73, 74, 82, 83, 84, 86, 87, 92, 106, 145, 170, 171; IV: 56-60.
 Advena Group, **73**, 75, 91, 92.
 Advena Section, **73**, *28*.
 Advena Stock, 13.
 alleni, **69-73**, *28*, 12, 13, 15, 19, 20, 21, 25, 35, 45, 74, 106, 108, 109, 113, 170, 171; IV: 51-55.
 Alleni Group, **69**, *28*, 35.
 Alleni Stock, 13, 71.
 apalachicolae, 55-58, *29*, 12, 13, 14, 20, 34, 35, 36, 37, 38, 51, 54, 59, 63, 66, 118, 170, 172; III: 26-30, XVII.

PROCAMBARUS, (cont.)
 barbatus, **39-41**, 8, 9, 34, 35, 36, 37, 38, 44, 48, 54, 70, 108, 109; II: 1-5.
 Barbatus Group, **35**, *29*, 27, 37, 38, 48, 49, 52, 58, 60, 61, 63, 65, 68, 69.
 Barbatus Section, **33**, *28*, 35, 63, 71.
 Barbatus Stock, 13, 100.
 Barbatus Subgroup, **38**, 34, 35, 39, 60.
 bivittatus, 96-100, *31*, 12, 14, 20, 21, 63, 94, 95, 103, 130, 170, 171; VI: 96-100, XXI.
 blandingii, 63, 95, 157.
 blandingii acutus, **94-96**, *31*, 14, 20, 21, 64, 74, 98, 103, 109, 169, 170, 171; VI: 91-95.
 blandingii blandingii, 94.
 blandingii cuevachicae, 94.
 Blandingii Group, **93**, *30*, *31*, 98, 109.
 Blandingii Section, **93**, *28*, 37, 63, 94, 100, 109.
 Blandingii Subgroup, **93-94**, *31*, 115.
 clarkii, 8, 9, 99, 100, 103, 104, 105.
 clarkii clarkii, 104.
 Clarkii Group, 98.
 clarkii paeninsulanus, 104.
 Clarkii Subgroup, **98-100**, *30*, 93, 105, 115.
 Clarkii Stock, 100.
 econfinae, 49-52, *29*, 12, 13, 20, 21, 34, 35, 36, 37, 38, 48, 57, 58, 63, 170; II: 16-20, XIV.
 escambiensis, 46-49, *29*, 12, 13, 20, 34, 35, 36, 37, 38, 39, 40, 44, 51, 59, 63, 68, 69, 170; II: 11-15, XIV.
 evermanni, **107-110**, *32*, 13, 14, 21, 70, 93, 95, 112, 115, 170, 171; VII, 111-115.
 Evermanni Subgroup, **107**, *31*, 93.
 fallax, **111-114**, *31*, 15, 18, 20, 21, 45, 71, 83, 106, 109, 110, 115, 116, 117, 118, 123, 124, 128, 145, 152, 153, 170, 171; VII, 116-120.
 Fallax Subgroup, **111**, *31*, 93, 115, 117.
 geodytes, 80-83, *30*, 12, 13, 20, 23, 73, 74, 75, 171; V: 61-65, XIX.
 gracilis, 35, 71, 109.
 hagenianus, 35, 71, 109.
 hayi, 63, 94, 95, 109.
 hubbelli, **67-69**, *30*, 9, 12, 14, 15, 20, 21, 34, 35, 36, 38, 60, 106, 118, 155, 156, 166, 170, 171, 172; IV: 46-50.
 Hubbelli Subgroup, **67**, 35, 60.
 kilbyi, **64-67**, *29*, 12, 13, 14, 15, 20, 21, 34, 35, 36, 37, 38, 55, 69, 87, 106, 116, 118, 170, 171, 172; IV: 41-45.
 Kilbyi Subgroup, **64**, 35, 60.
 latipleurum, 52-55, *29*, 12, 13, 14, 15, 20, 21, 28, 34, 35, 36, 37, 38, 39, 51, 57, 58, 63, 66, 87, 118, 170; III: 21-25, XVI.
 lecontei, 94, 95, 98.
 leonensis, **114-117**, *32*, 9, 15, 20, 21, 66, 87, 106, 110, 111, 112, 118, 123, 152, 155, 170, 171; VIII: 121-125.
 lucifugus, 131.
 lucifugus alachua, **136-139**, *31*, 12, 13, 21, 129, 130, 134, 135, 140, 141, 147, 170; IX: 156-160.

INDEX

PROCAMBARUS, (cont.)
lucifugus intergrades, **139**, 134, 171.
lucifugus lucifugus, **134-136**, *31*, 12, 13, 21, 129, 130, 137, 139, 147, 170, 171; IX: 151-155.
Lucifugus Subgroup, **134**, 93, 129.
lunzi, 129, 142, 144.
mexicanus, 38.
okaloosae, 100-104, *31*, 12, 14, 20, 21, 60, 63, 95, 96, 99, 105, 123, 128, 129, 152, 170, 171; VII: 101-105, XXII.
paeninsulanus, **104-107**, *31*, 14, 15, 20, 21, 25, 45, 66, 68, 71, 72, 87, 99, 100, 103, 109, 112, 113, 116, 118, 123, 128, 145, 155, 166, 170, 171, 172; VII: 106-110.
pallidus, **139-142**, *31*, 12, 21, 129, 130, 134, 135, 137, 138, 147, 170; X: 161-165.
pearsei, 41.
pictus, **130-131**, *31*, 10, 12, 13, 15, 21, 106, 129, 133, 134, 170; IX: 141-145.
Pictus Group, **129**, *30*, 12, 93, 115, 134, 140, 142, 147.
Pictus Stock, 13.
Pictus Subgroup, **129-130**, 93.
pubescens, 122, 129, 130, 133.
pubischelae, 41-46, *29*, *30*, 13, 15, 20, 21, 23, 34, 35, 36, 37, 38, 48, 54, 59, 63, 69, 71, 78, 87, 106, 113, 145, 170, 171; II: 6-10, XIII.
pycnogonopodus, **117-119**, *32*, 9, 12, 14, 15, 20, 21, 52, 55, 58, 66, 68, 87, 106, 110, 111, 112, 115, 116, 123, 128, 152, 166, 170, 171, 172; VIII: 126-130.
pygmaeus, 83-88, *30*, 13, 14, 15, 20, 55, 66, 73, 74, 75, 145, 170, 171; V: 66-67, XX.
rathbunae, **59-60**, *30*, 12, 15, 20, 21, 34, 35, 36, 37, 38, 54, 57, 63, 69, 103, 170, 171; III: 31-35.
rogersi, 9, 13, 15, 60, 86, 88, 90.
rogersi campestris, **90**, *30*, 12, 14, 20, 66, 73, 74, 88, 172; VI: 81-85.
Rogersi Group, **88**, 73.

PROCAMBARUS, (cont.)
rogersi intergrades, **90-91**, 58, 74, 170, 171.
rogersi ochlocknensis, **89-90**, *30*, 12, 14, 15, 20, 66, 73, 74, 88, 170, 171; V: 76-80.
rogersi rogersi, **89**, *30*, 12, 14, 15, 20, 58, 66, 73, 74, 88, 90, 170; V: 71-75.
Rogersi Stock, 87.
seminolae, 142-146, *31*, 13, 15, 20, 21, 45, 78, 87, 106, 113, 124, 129, 130, 170, 171; X: 166-170, XXIV.
Seminolae Subgroup, **142**, 93, 129.
shermani, 61-64, *29*, 12, 14, 20, 33, 34, 35, 36, 37, 38, 48, 60, 96, 169, 170, 171; III: 36-40, XVIII.
Shermani Subgroup, **60**, 35.
simulans, 35, 70, 71, 109.
spiculifer, **119-125**, *32*, 13, 14, 15, 20, 21, 66, 103, 106, 113, 116, 118, 127, 128, 145, 152, 153, 155, 162, 166, 170, 171, 172; VIII: 131-135.
Spiculifer Group, **119**, *30*, 91, 93.
troglodytes, 99, 100, 105.
versutus, **126-129**, *32*, 13, 14, 15, 20, 21, 92, 95, 103, 109, 118, 119, 120, 121, 123, 166, 170, 171, 172; VIII: 136-140.
wiegmanni, 70, 109.
youngi, 131-134, *31*, 12, 13, 14, 15, 87, 129, 130, 170; IX: 146-150, XXIII.
Youngi Stock, 87.
Seepage Areas, 19.
spiculifer, Astacus, 119.
Subterranean Situations, see Ecology.

Taxonomic Characters, evaluation of: Areola, 25; Chela, 25; Female, 25; First Pleopod of male, 25; Hooks on ischiopodites, 23; Spines, 25.
Terminology and Measurements, 24.
TROGLOCAMBARUS, 146, *28*, 9, 11, 22, 23.
maclanei, **146-149**, *28*, 7, 9, 12, 13, 21, 136, 141, 170; I, X: 171-175.

Underground Waters, see Ecology.

Plate III

Vertical Rows: 1—Dorsal view of carapace. 2—Upper surface of chela of first form male. 3—Annulus ventralis of female. 4—Lateral view of first pleopod of second form male. 5—Lateral view of first pleopod of first form male.

Figs. 21-25, *Procambarus latipleurum*
Figs. 26-30, *Procambarus apalachicolae*
Figs. 31-35, *Procambarus rathbunae*
Figs. 36-40, *Procambarus shermani*

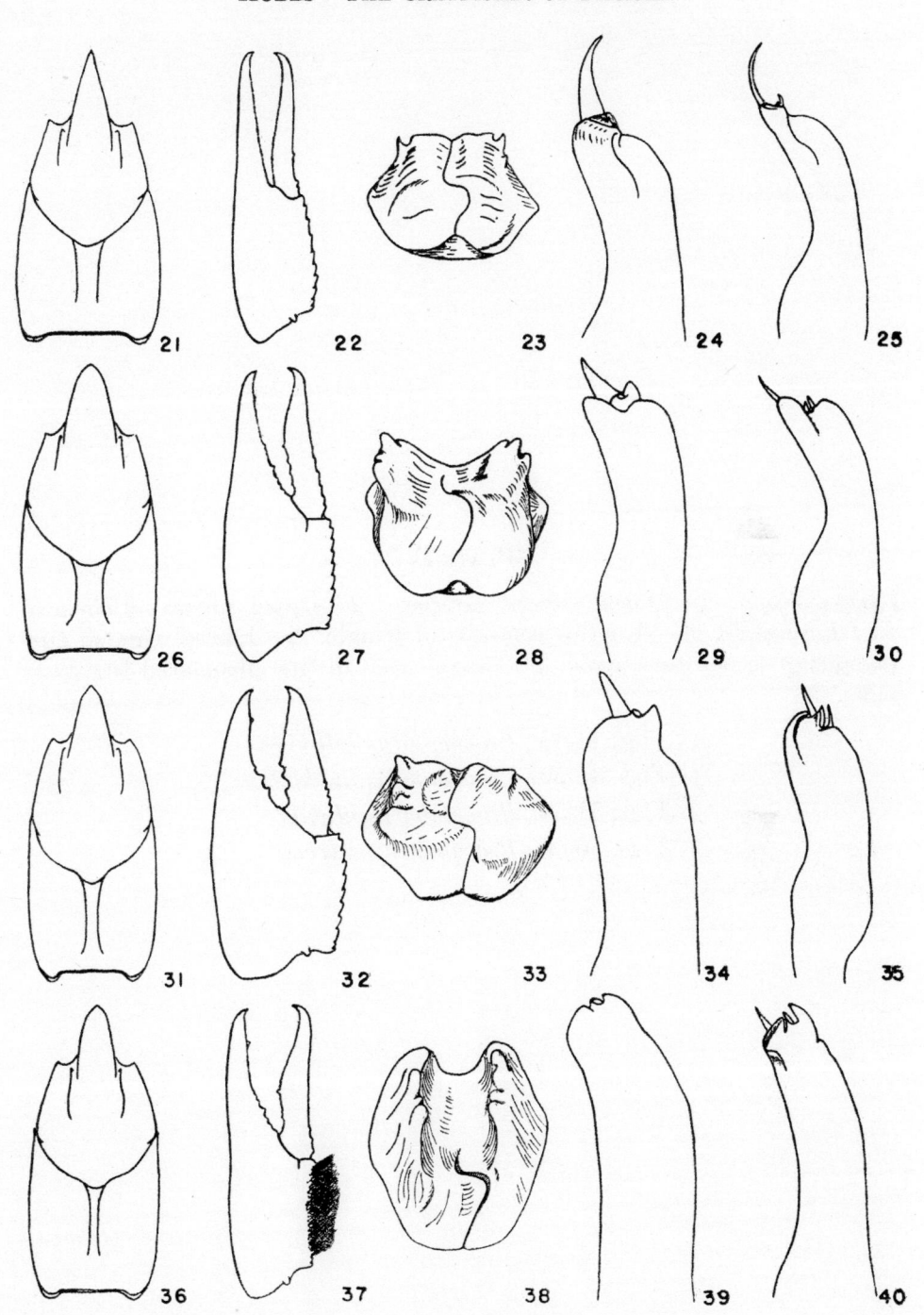

PLATE III

Plate IV

Vertical Rows: 1—Dorsal view of carapace. 2—Upper surface of chela of first form male. 3—Annulus ventralis of female. 4—Lateral view of first pleopod of second form male. 5—Lateral view of first pleopod of first form male.

Figs. 41-45, *Procambarus kilbyi*
Figs. 46-50, *Procambarus hubbelli*
Figs. 51-55, *Procambarus alleni*
Figs. 56-60, *Procambarus advena*

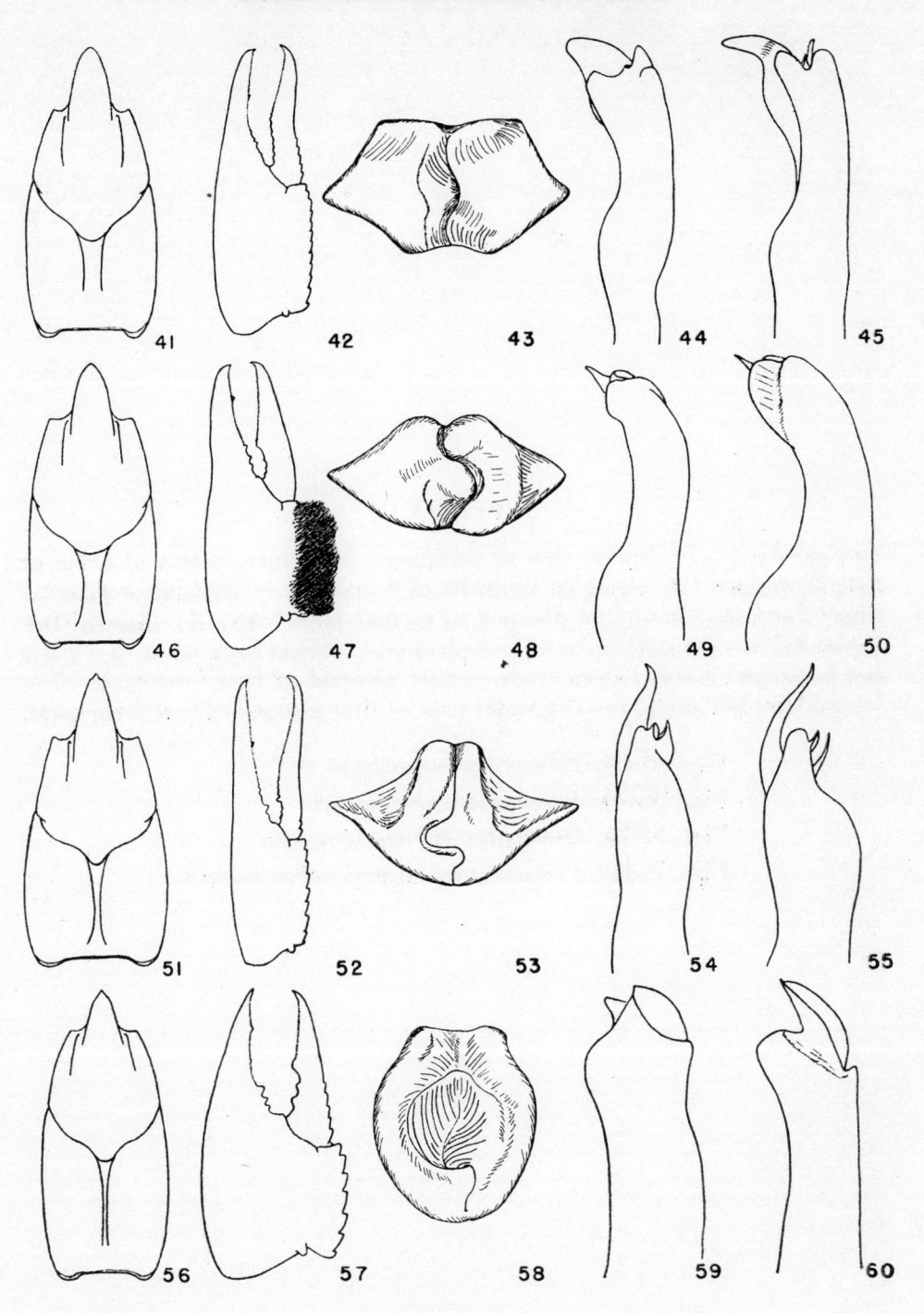

PLATE IV

Plate V

Vertical Rows: 1—Dorsal view of carapace. 2—Upper surface of chela of first form male. 3—Annulus ventralis of female. 4—(first two horizontal rows) Lateral view of first pleopod of second form male. 4—(second two horizontal rows) Caudal view of first pleopod of second form male. 5—(first two horizontal rows) Lateral view of first pleopod of first form male. 5—(second two horizontal rows) Caudal view of first pleopod of first form male.

Figs. 61-65, *Procambarus geodytes*
Figs. 66-70, *Procambarus pygmaeus*
Figs. 71-75, *Procambarus rogersi rogersi*
Figs. 76-80, *Procambarus rogersi ochlocknensis*

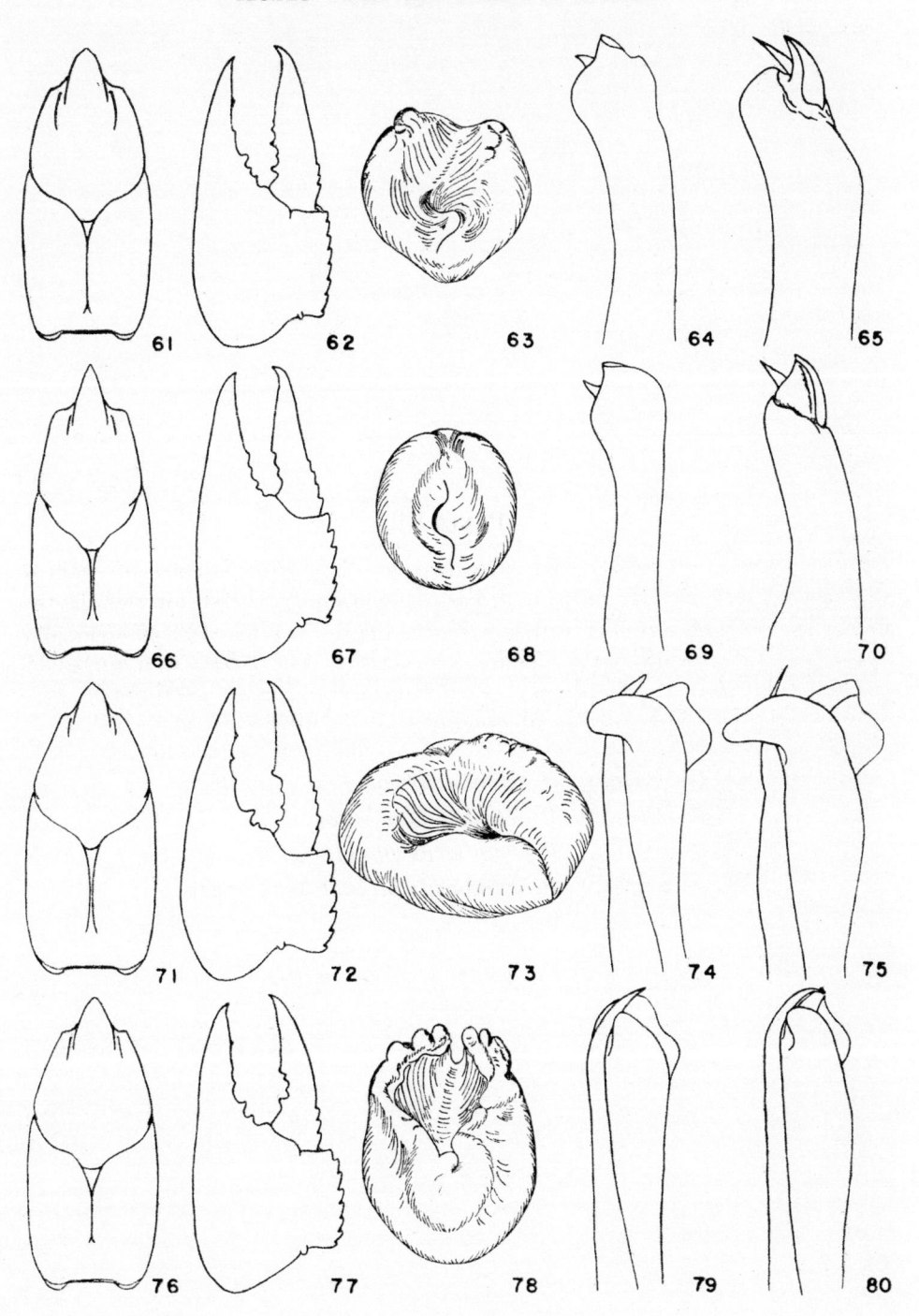

PLATE V

Plate VII

Vertical Rows: 1—Dorsal view of carapace. 2—Upper surface of chela of first form male. 3—Annulus ventralis of female. 4—Lateral view of first pleopod of second form male. 5—Lateral view of first pleopod of first form male.

Figs. 101-105, *Procambarus okaloosae*
Figs. 106-110, *Procambarus paeninsulanus*
Figs. 111-115, *Procambarus evermanni*
Figs. 116-120, *Procambarus fallax*

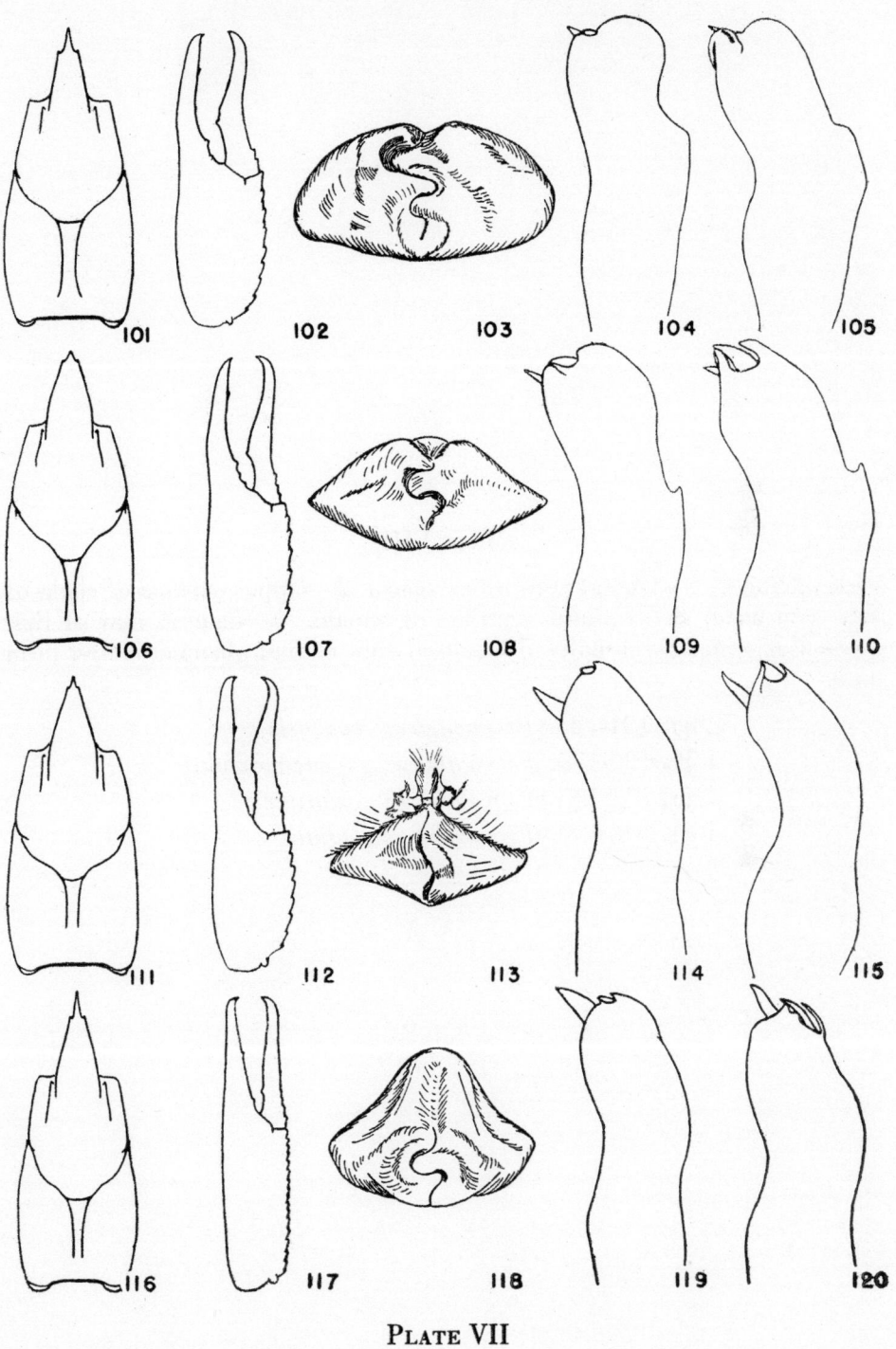

PLATE VII

PLATE VIII

Vertical Rows: 1—Dorsal view of carapace. 2—Upper surface of chela of first form male. 3—Annulus ventralis of female. 4—Lateral view of first pleopod of second form male. 5—Lateral view of first pleopod of first form male.

Figs. 121-125, *Procambarus leonensis*
Figs. 126-130, *Procambarus pycnogonopodus*
Figs. 131-135, *Procambarus spiculifer*
Figs. 136-140, *Procambarus versutus*

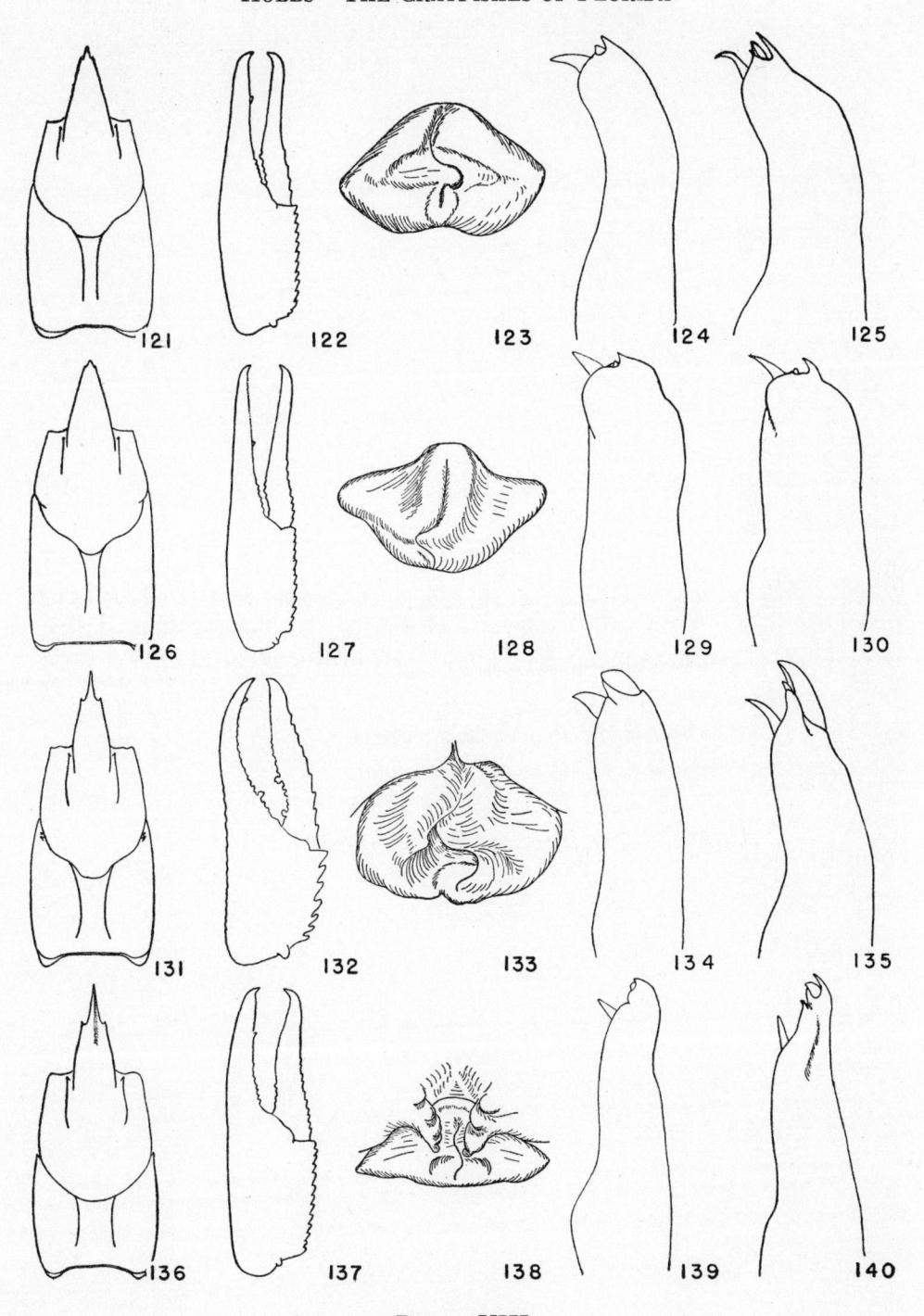

PLATE VIII

Plate IX

Vertical Rows: 1—Dorsal view of carapace. 2—Upper surface of chela of first form male. 3—Annulus ventralis of female. 3—Lateral view of first pleopod of second form male. 5—Lateral view of first pleopod of first form male.

        Figs. 141-145, *Procambarus pictus*
        Figs. 146-150, *Procambarus youngi*
        Figs. 151-155, *Procambarus lucifugus lucifugus*
        Figs. 156-160, *Procambarus lucifugus alachua*

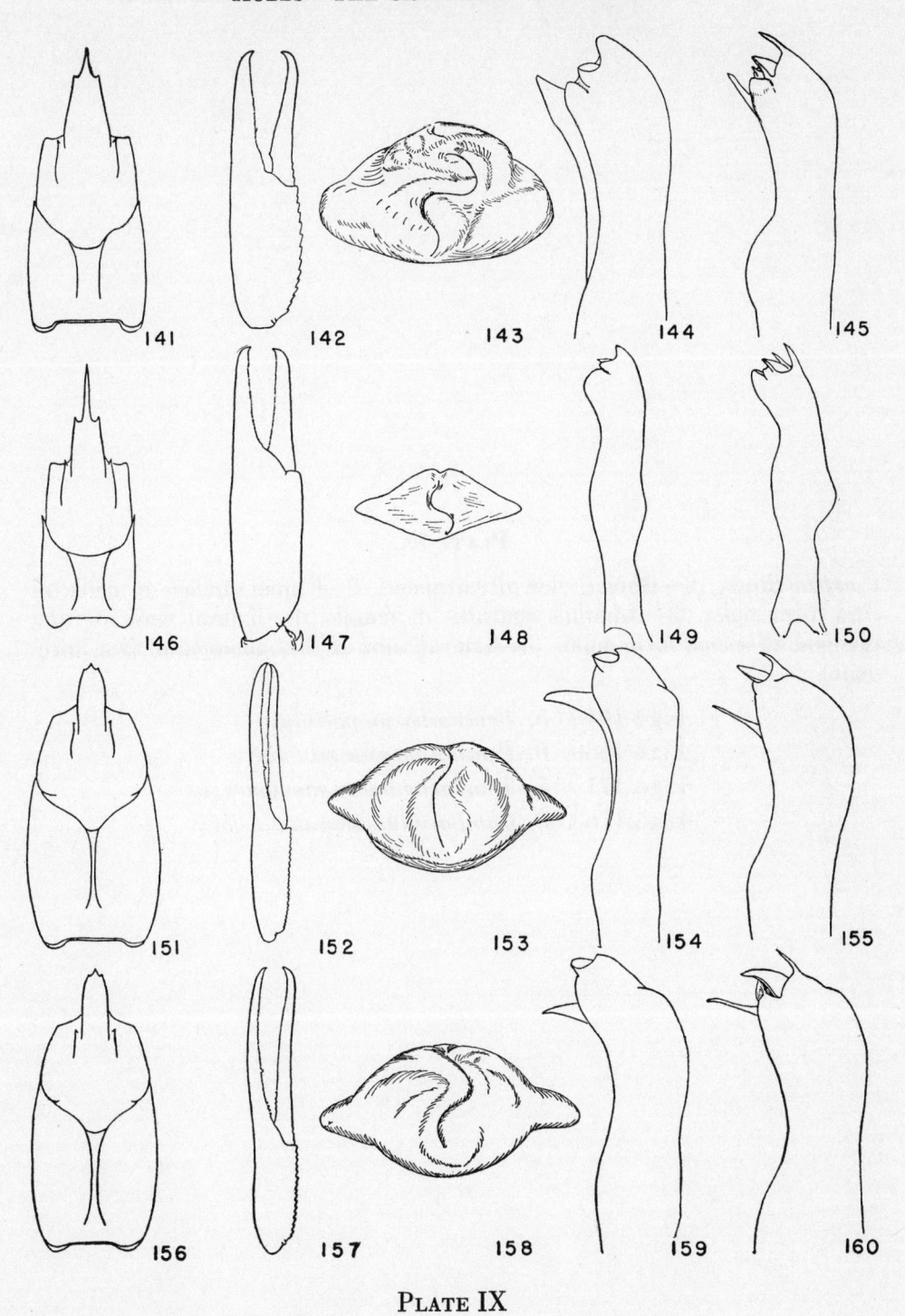

PLATE IX

PLATE X

Vertical Rows: 1—Dorsal view of carapace. 2—Upper surface of chela of first form male. 3—Annulus ventralis of female. 4—Lateral view of first pleopod of second form male. 5—Lateral view of first pleopod of first form male.

Figs. 161-165, *Procambarus pallidus*
Figs. 166-170, *Procambarus seminolae*
Figs. 171-175, *Troglocambarus maclanei*
Figs. 176-180, *Cambarellus schmitti*

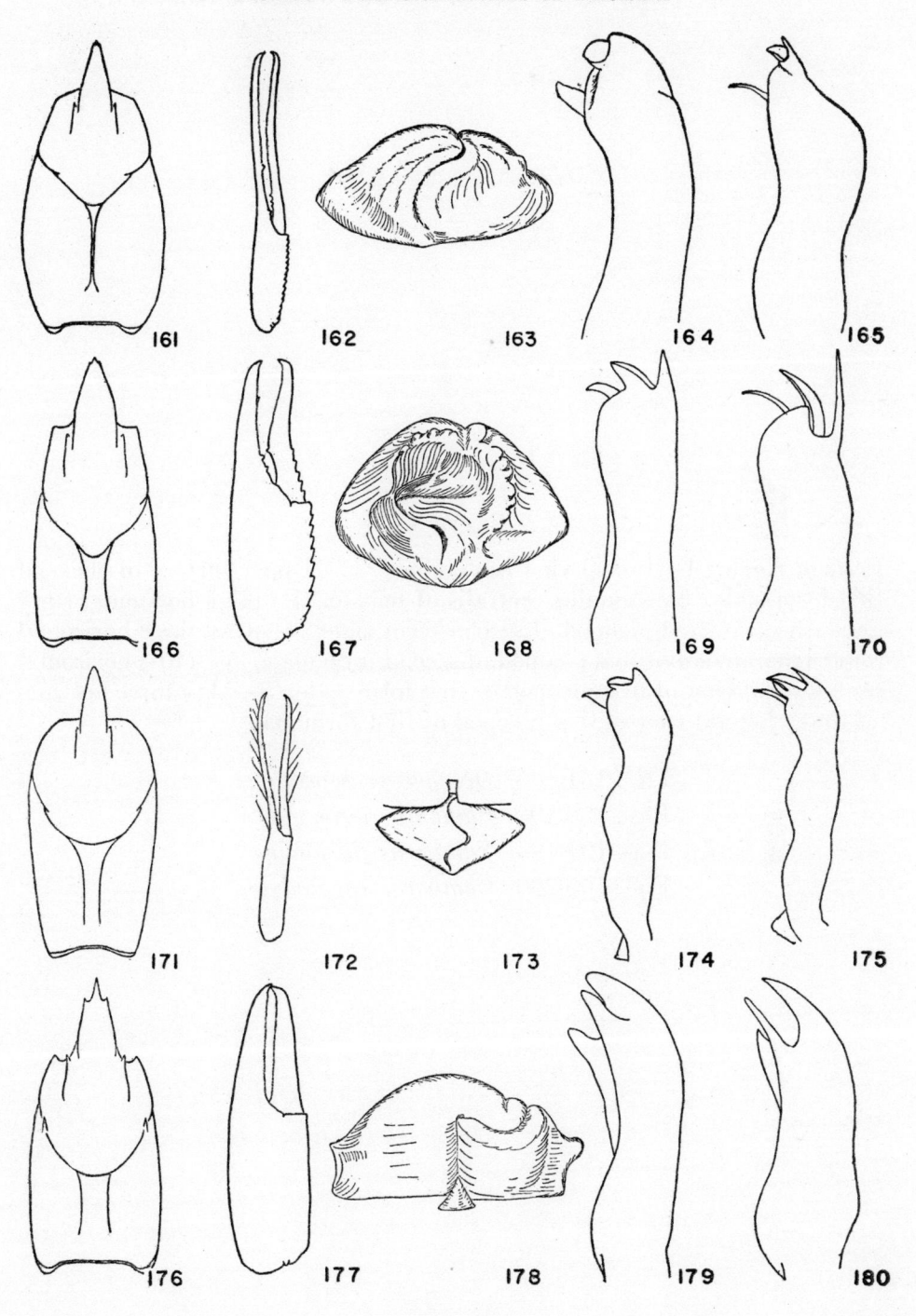

PLATE X

Plate XI

Vertical Rows: 1—Dorsal view of carapace. 2—Upper surface of chela of first form male. 3—Annulus ventralis of female. 4—(first horizontal row) Caudal view of first pleopod of second form male. 4—(last three horizontal rows) Lateral view of first pleopod of second form male. 5—(first horizontal row) Caudal view of first pleopod of first form male. 5—(last three horizontal rows) Lateral view of first pleopod of first form male.

        Figs. 181-185, *Orconectes clypeata*
        Figs. 186-190, *Cambarus latimanus*
        Figs. 191-195, *Cambarus floridanus*
        Figs. 196-200, *Cambarus cryptodytes*

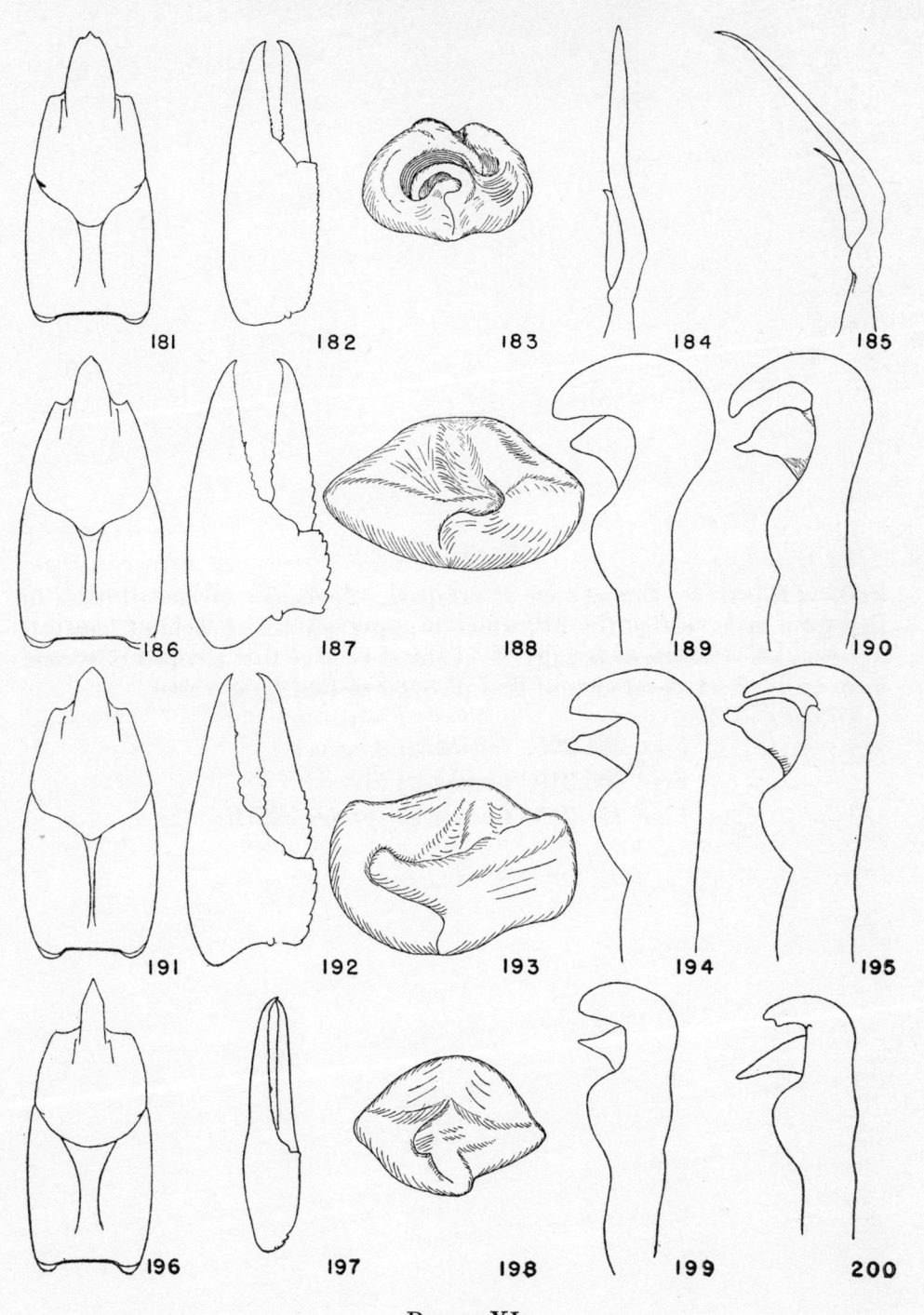

PLATE XI

Plate XII

Vertical Rows: 1—Dorsal view of carapace. 2—Upper surface of chela of first form male (except fig. 212 which is upper surface of chela of female). 3—Annulus ventralis of female. 4—Lateral view of first pleopod of second form male. 5—Lateral view of first pleopod of first form male.

>Figs. 201-205, *Cambarus diogenes*
>Figs. 206-210, *Cambarus byersi*
>Figs. 211-215, *Cambarus species incertis*

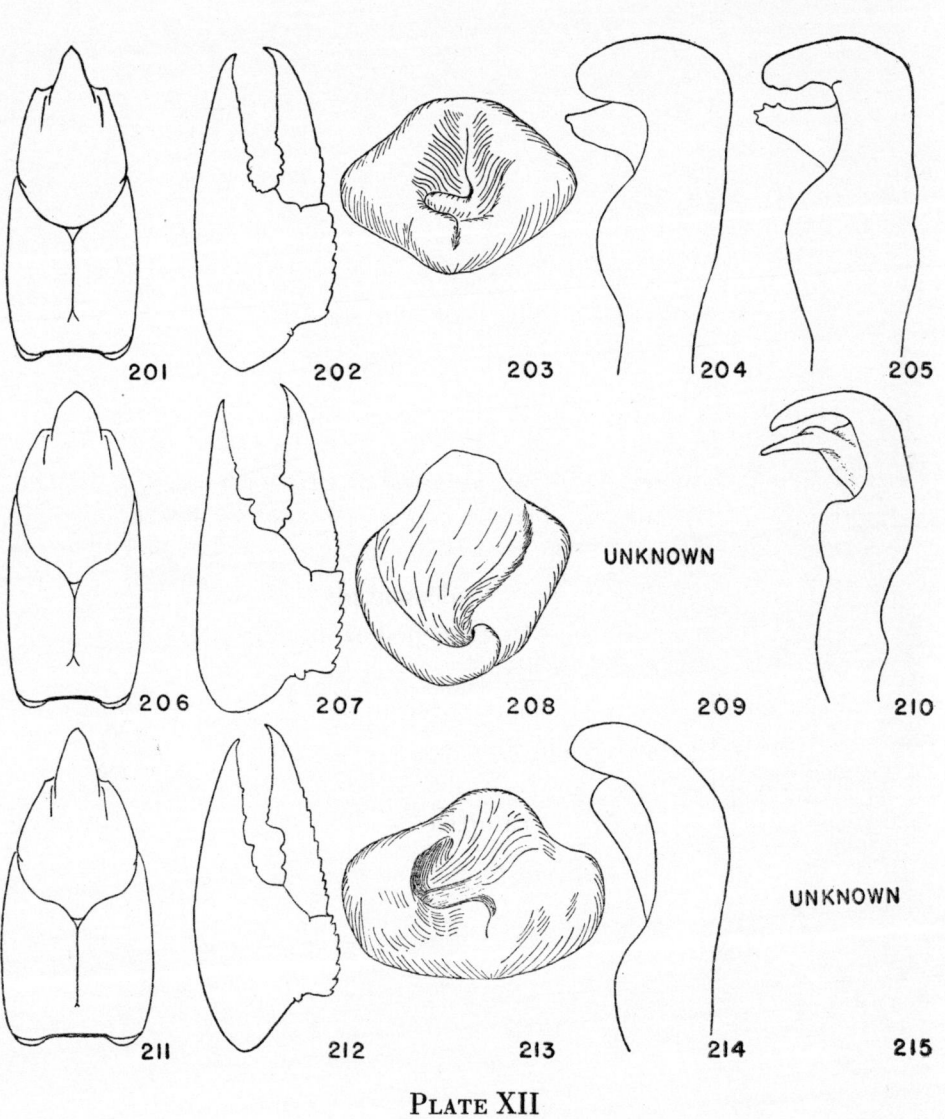

PLATE XII

EXPLANATION OF PLATE XVI

Procambarus latipleurum

(Pubescence has been removed from all structures illustrated)

Fig. 246, Mesial view of first pleopod of first form male
Fig. 247, Mesial view of first pleopod of second form male
Fig. 248, Epistome
Fig. 249, Lateral view of carapace
Fig. 250, Antennal scale
Fig. 251, Lateral view of first pleopod of second form male
Fig. 252, Lateral view of first pleopod of first form male
Fig. 253, Hook on ischiopodite of third pereiopod
Fig. 254, Hook on ischiopodite of fourth pereiopod
Fig. 255, Annulus ventralis

PLATE XVI

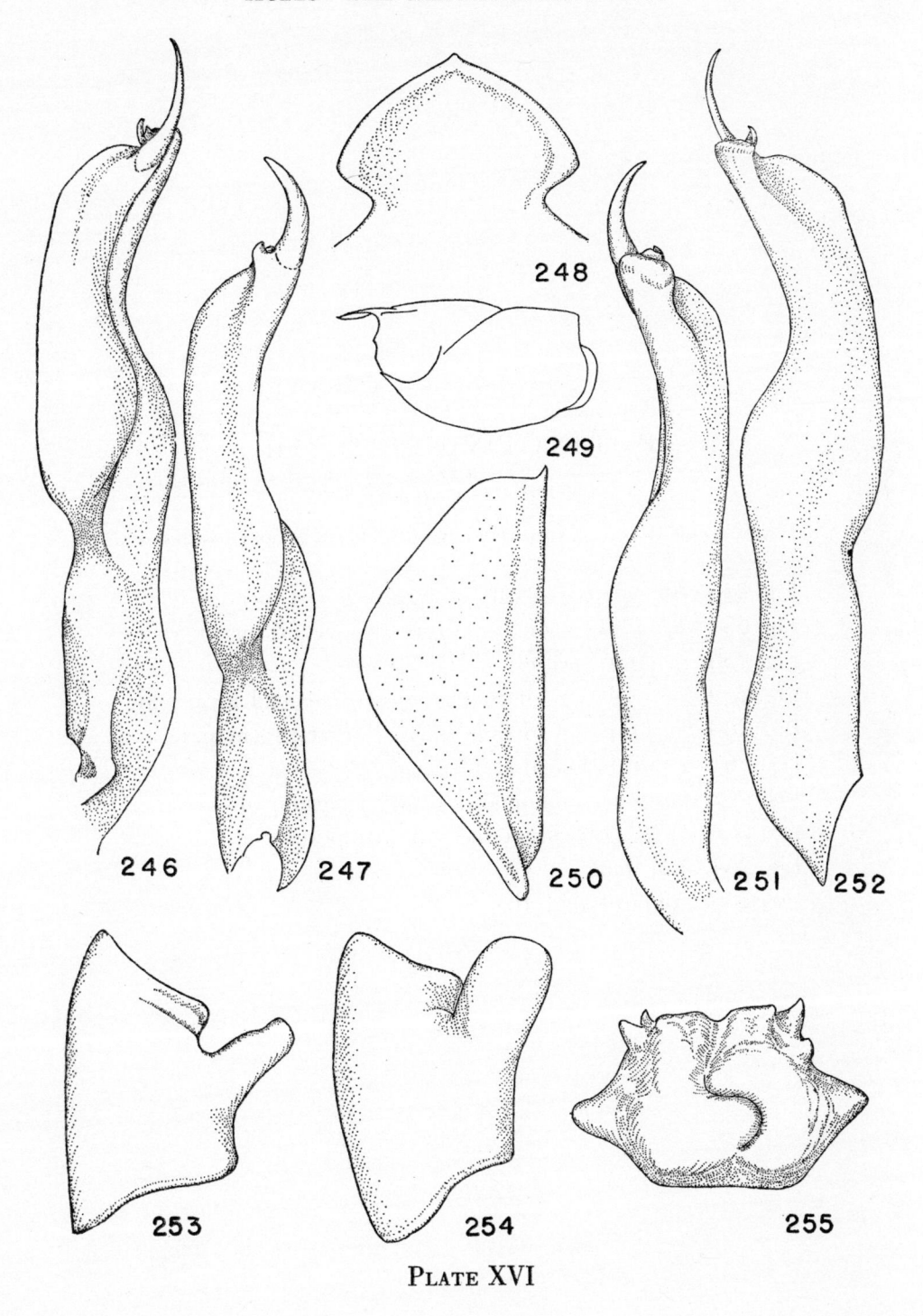

EXPLANATION OF PLATE XVIII

Procambarus shermani

(Pubescence has been removed from all structures illustrated)

Fig. 266, Mesial view of first pleopod of first form male
Fig. 267, Mesial view of first pleopod of second form male
Fig. 268, Lateral view of carapace
Fig. 269, Epistome
Fig. 270, Antennal scale
Fig. 271, Lateral view of first pleopod of second form male
Fig. 272, Lateral view of first pleopod of first form male
Fig. 273, Hook on ischiopodite of third pereiopod
Fig. 274, Hook on ischiopodite of fourth pereiopod
Fig. 275, Annulus ventralis

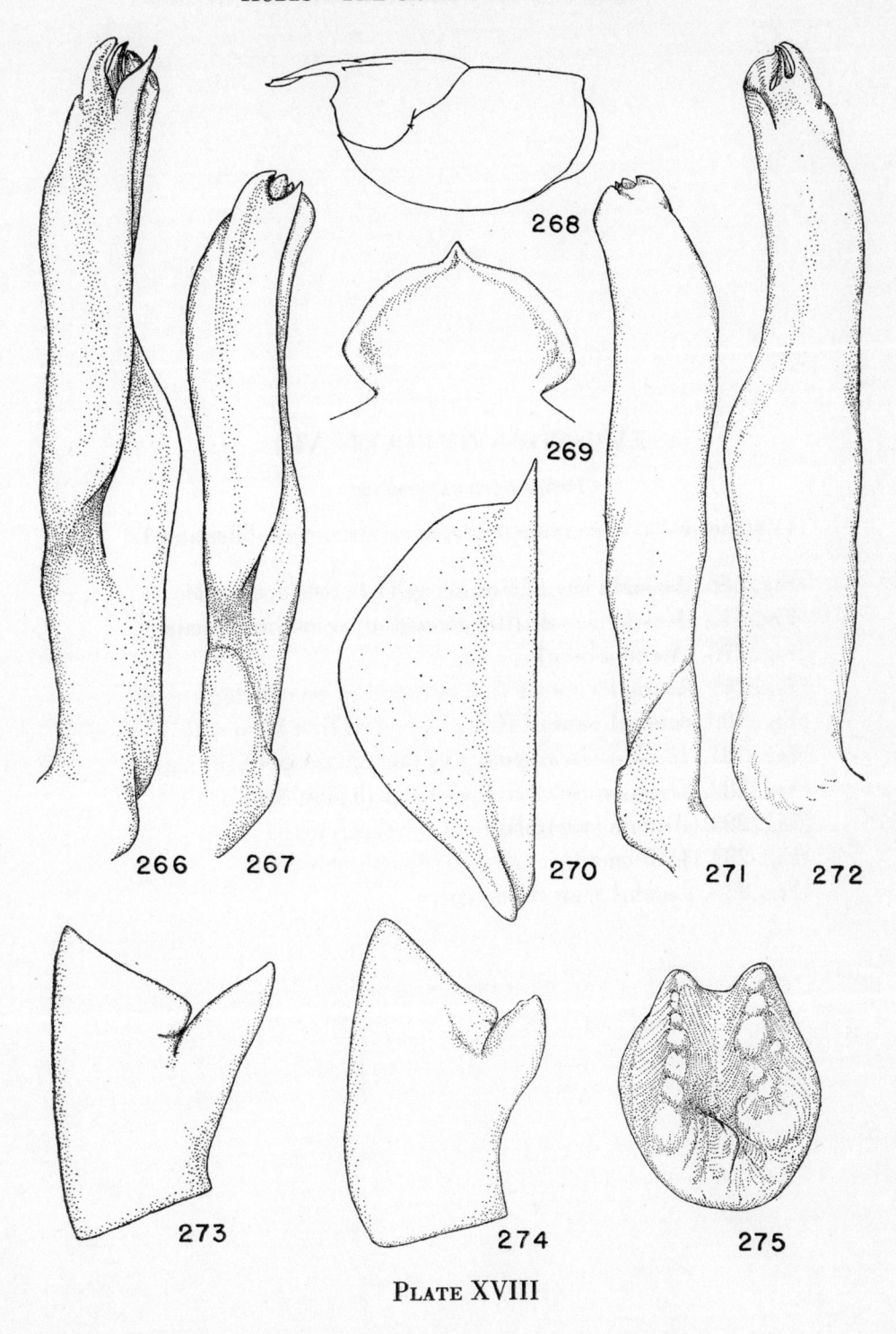

PLATE XVIII

EXPLANATION OF PLATE XIX

Procambarus geodytes

(Pubescense has been removed from all structures illustrated)

Fig. 276, Mesial view of first pleopod of first form male
Fig. 277, Mesial view of first pleopod of second form male
Fig. 278, Antennal scale
Fig. 279, Lateral view of first pleopod of second form male
Fig. 280, Lateral view of first pleopod of first form male
Fig. 281, Hook on ischiopodite of third pereiopod
Fig. 282, Hook on ischiopodite of fourth pereiopod
Fig. 283, Annulus ventralis
Fig. 284, Epistome
Fig. 285, Lateral view of carapace

Hobbs—The Crayfishes of Florida

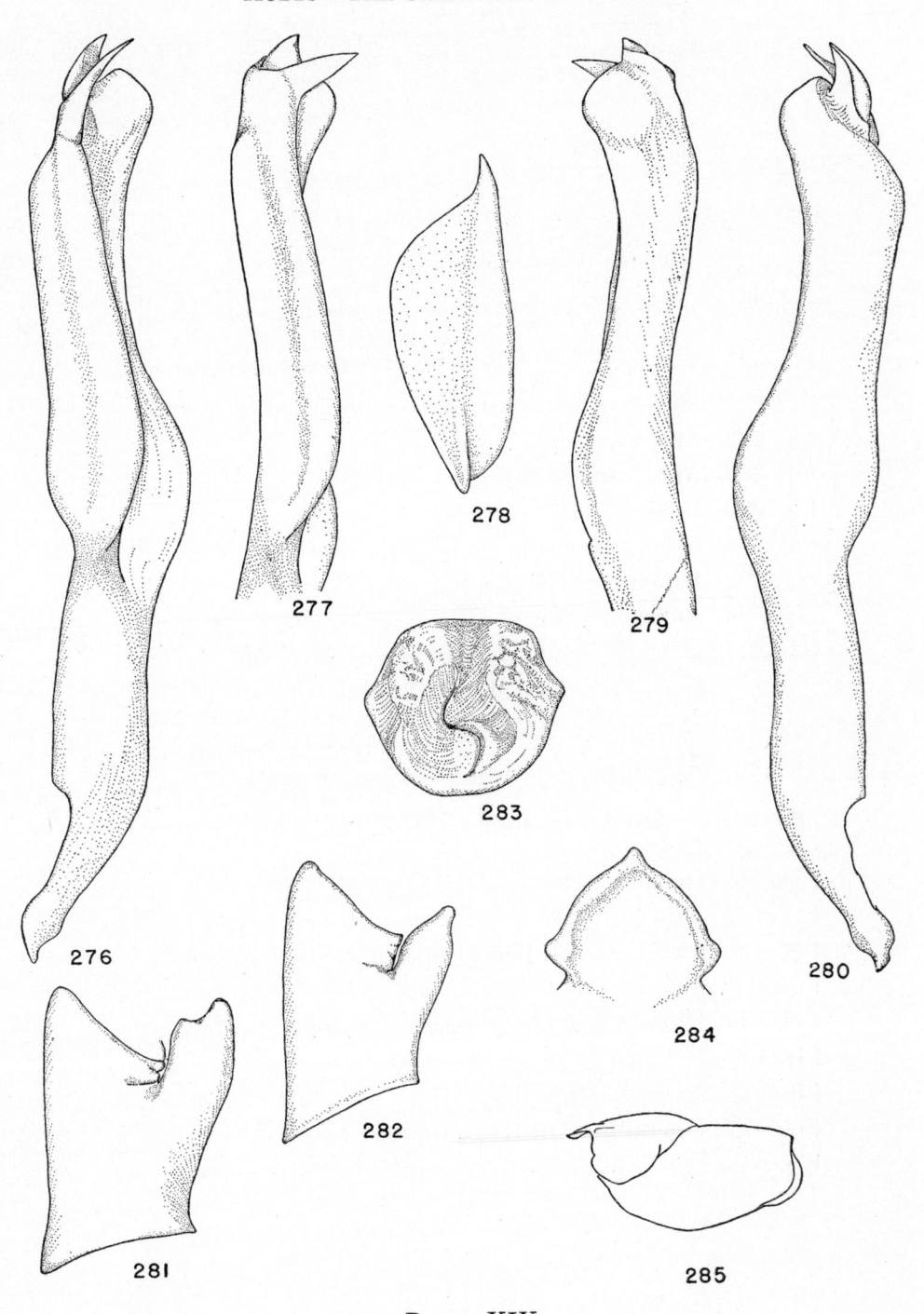

Plate XIX

EXPLANATION OF PLATE XX

Cambarellus schmitti (Figs. 286-295)

(Pubescence has been removed from all structures illustrated)

Fig. 286, Mesial view of first pleopod of first form male
Fig. 287, Mesial view of first pleopod of second form male
Fig. 288, Epistome
Fig. 289, Annulus ventralis
Fig. 290, Lateral view of carapace
Fig. 291, Lateral view of first pleopod of second form male
Fig. 292, Lateral view of first pleopod of first form male
Fig. 293, Antennal scale
Fig. 294, Hook on ischiopodite of second pereiopod
Fig. 295, Hook on ischiopodite of third pereiopod

Procambarus pygmaeus (Figs. 296-304)

(Pubescence has been removed from all structures illustrated)

Fig. 296, Hook on ischiopodite of third pereiopod
Fig. 297, Antennal scale
Fig. 298, Mesial view of first pleopod of first form male
Fig. 299, Mesial view of first pleopod of second form male
Fig. 300, Lateral view of carapace
Fig. 301, Epistome
Fig. 302, Annulus ventralis
Fig. 303, Lateral view of first pleopod of second form male
Fig. 304, Lateral view of first pleopod of first form male

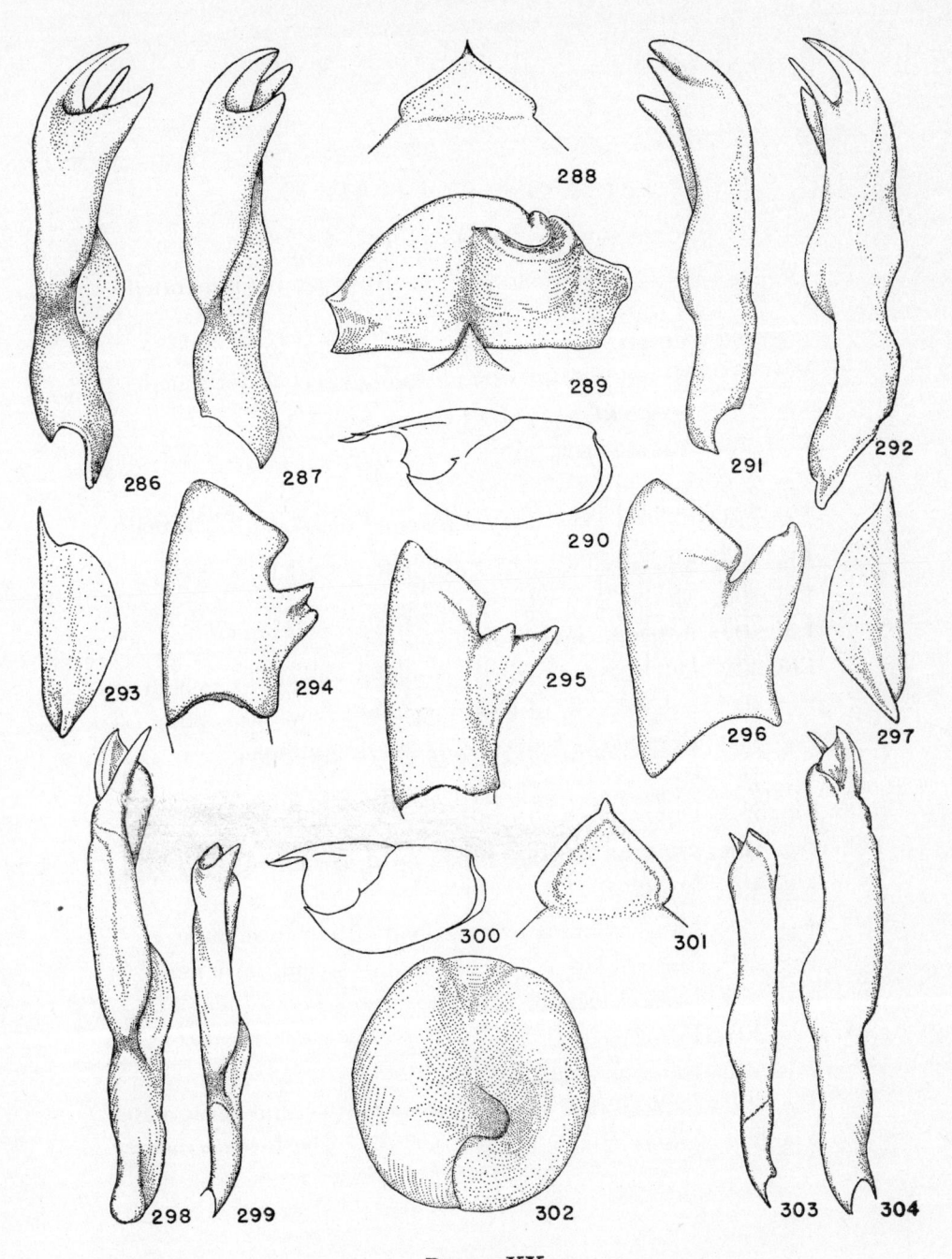

PLATE XX

EXPLANATION OF PLATE XXI

Procambarus bivittatus

(Pubescence has been removed from all structures illustrated)

Fig. 305, Mesial view of first pleopod of first form male
Fig. 306, Mesial view of first pleopod of second form male
Fig. 307, Antennal scale
Fig. 308, Lateral view of first pleopod of second form male
Fig. 309, Lateral view of first pleopod of first form male
Fig. 310, Hook on ischiopodite of third pereiopod
Fig. 311, Hook on ischiopodite of fourth pereiopod
Fig. 312, Lateral view of carapace
Fig. 313, Annulus ventralis
Fig. 314, Epistome

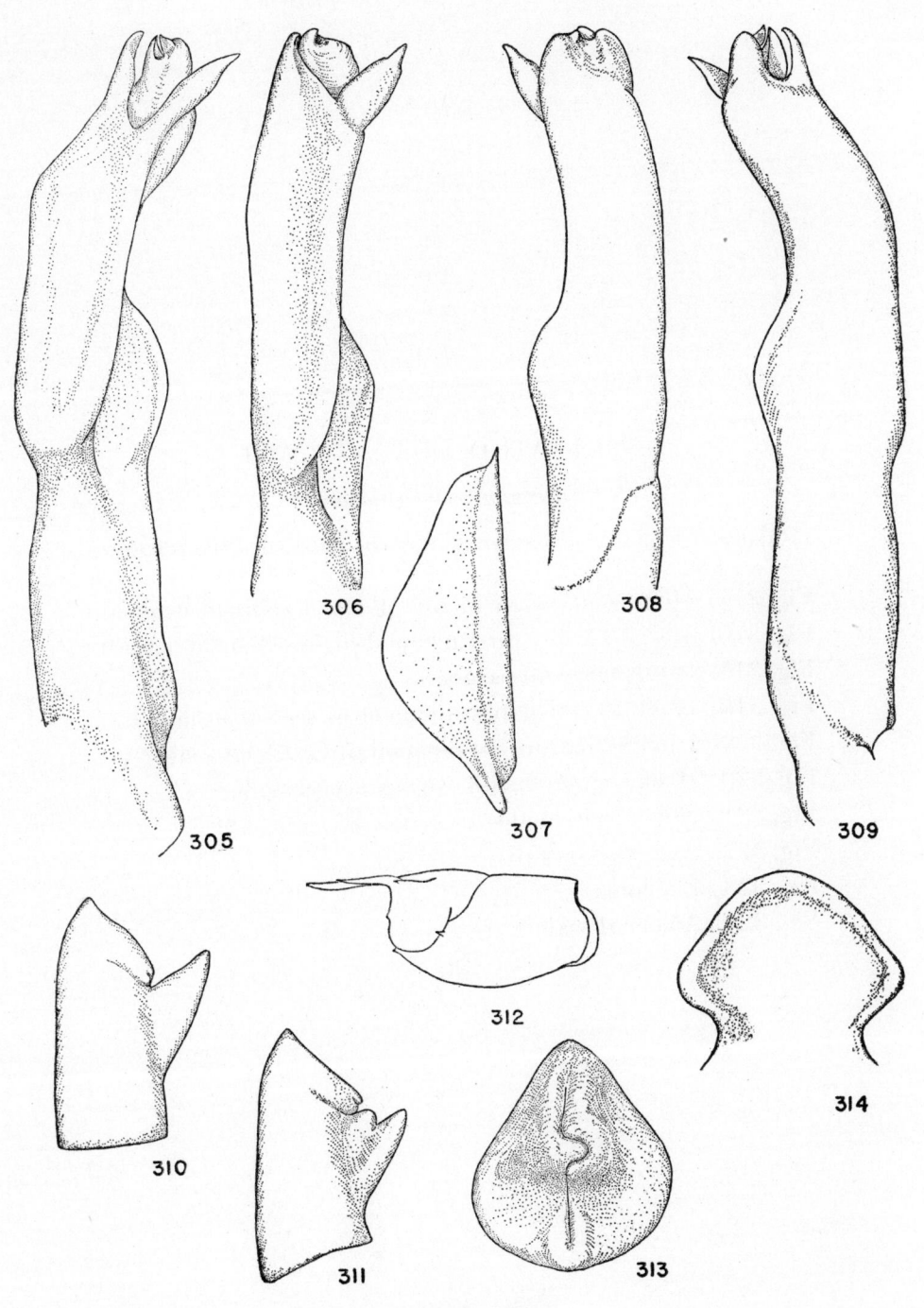

PLATE XXI

EXPLANATION OF PLATE XXII
Procambarus okaloosae

(Pubescence has been removed from all structures illustrated)

Fig. 315, Mesiocaudal view of first pleopod of first form male
Fig. 316, Mesial view of first pleopod of second form male
Fig. 317, Lateral view of carapace
Fig. 318, Lateral view of first pleopod of second form male
Fig. 319, Lateral view of first pleopod of first form male
Fig. 320, Hook on ischiopodite of third pereiopod
Fig. 321, Hook on ischiopodite of fourth pereiopod
Fig. 322, Annulus ventralis
Fig. 323, Epistome
Fig. 324, Antennal scale

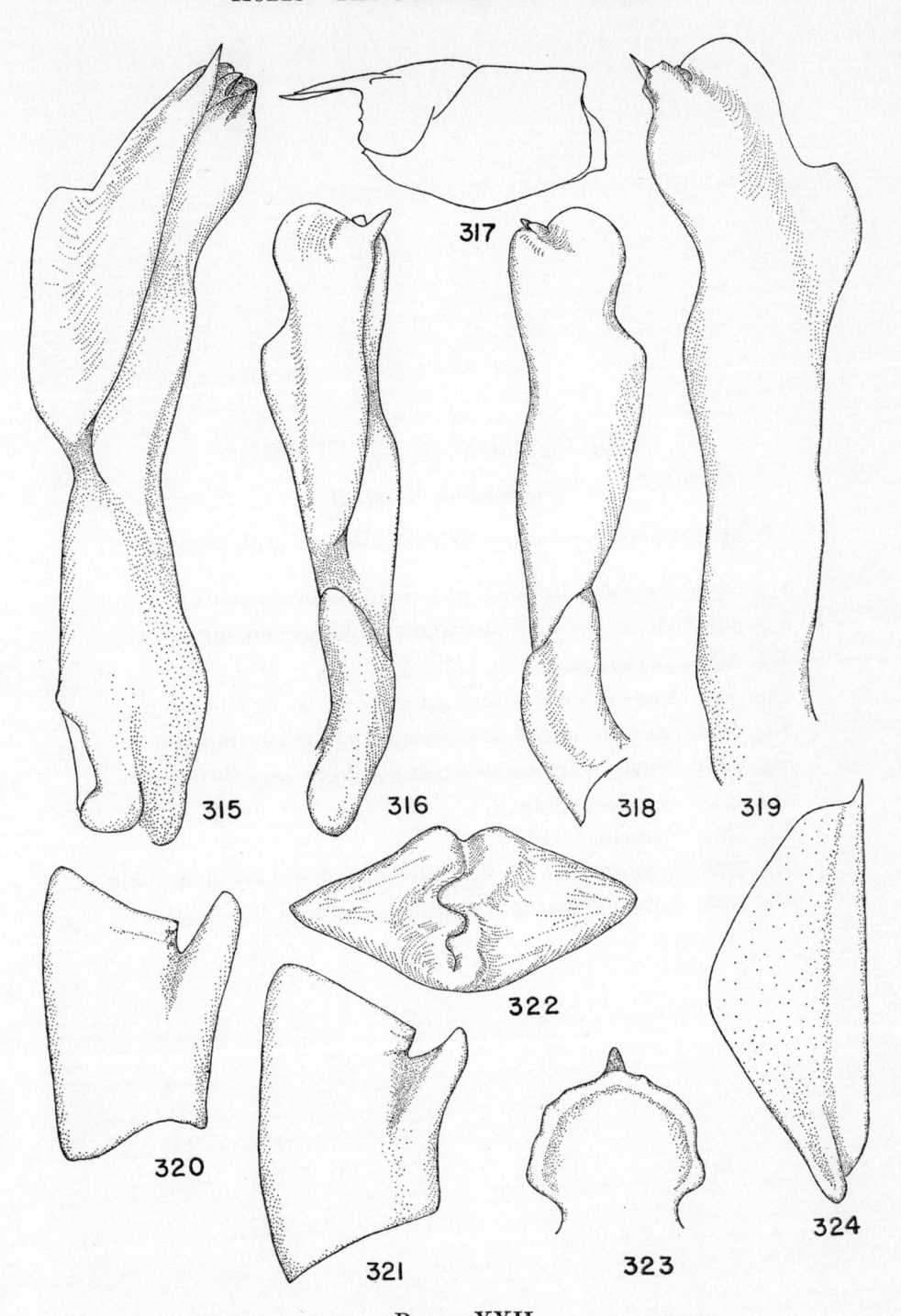

PLATE XXII

EXPLANATION OF PLATE XXIII

Procambarus youngi

(Pubescence has been removed from all structures illustrated)

Fig. 325, Hook on ischiopodite of third pereiopod
Fig. 326, Hook on ischiopodite of fourth pereiopod
Fig. 327, Epistome
Fig. 328, Lateral view of carapace
Fig. 329, Mesial view of first pleopod of first form male
Fig. 330, Mesial view of first pleopod of second form male
Fig. 331, Annulus ventralis
Fig. 332, Antennal scale
Fig. 333, Lateral view of first pleopod of second form male
Fig. 334, Lateral view of first pleopod of first form male

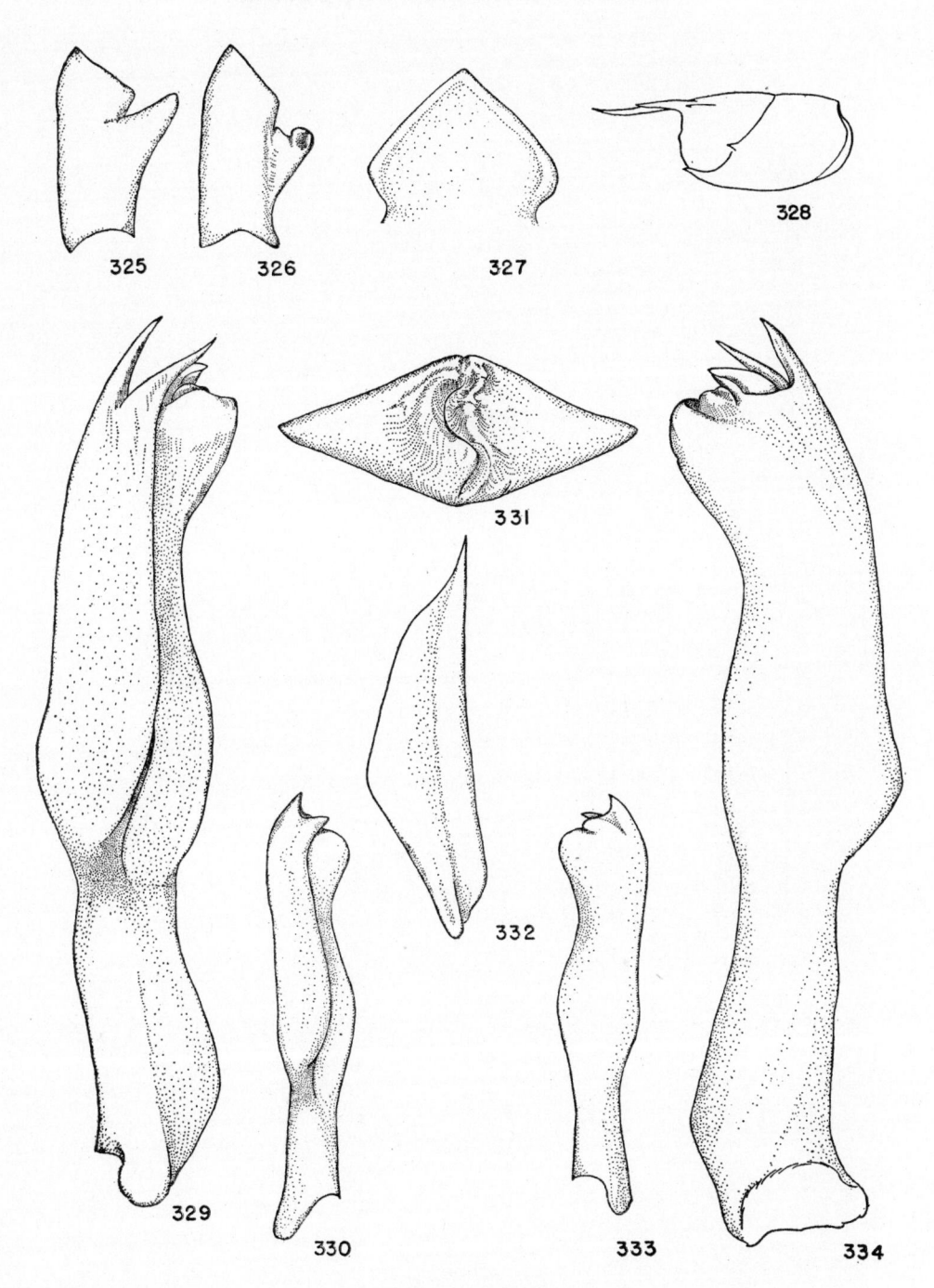

PLATE XXIII

EXPLANATION OF PLATE XXIV

Procambarus seminolae

(Pubescence has been removed from all structures illustrated)

Fig. 335, Mesial view of first pleopod of first form male
Fig. 336, Caudal view of first pleopod of first form male
Fig. 337, Cephalic view of first pleopod of first form male
Fig. 338, Lateral view of first pleopod of first form male
Fig. 339, Epistome
Fig. 340, Annulus ventralis
Fig. 341, Lateral view of carapace
Fig. 342, Mesial view of first pleopod of second form male
Fig. 343, Hook on ischiopodite of third pereiopod
Fig. 344, Hook on ischiopodite of fourth pereiopod
Fig. 345, Antennal scale
Fig. 346, Lateral view of first pleopod of second form male

Hobbs—The Crayfishes of Florida

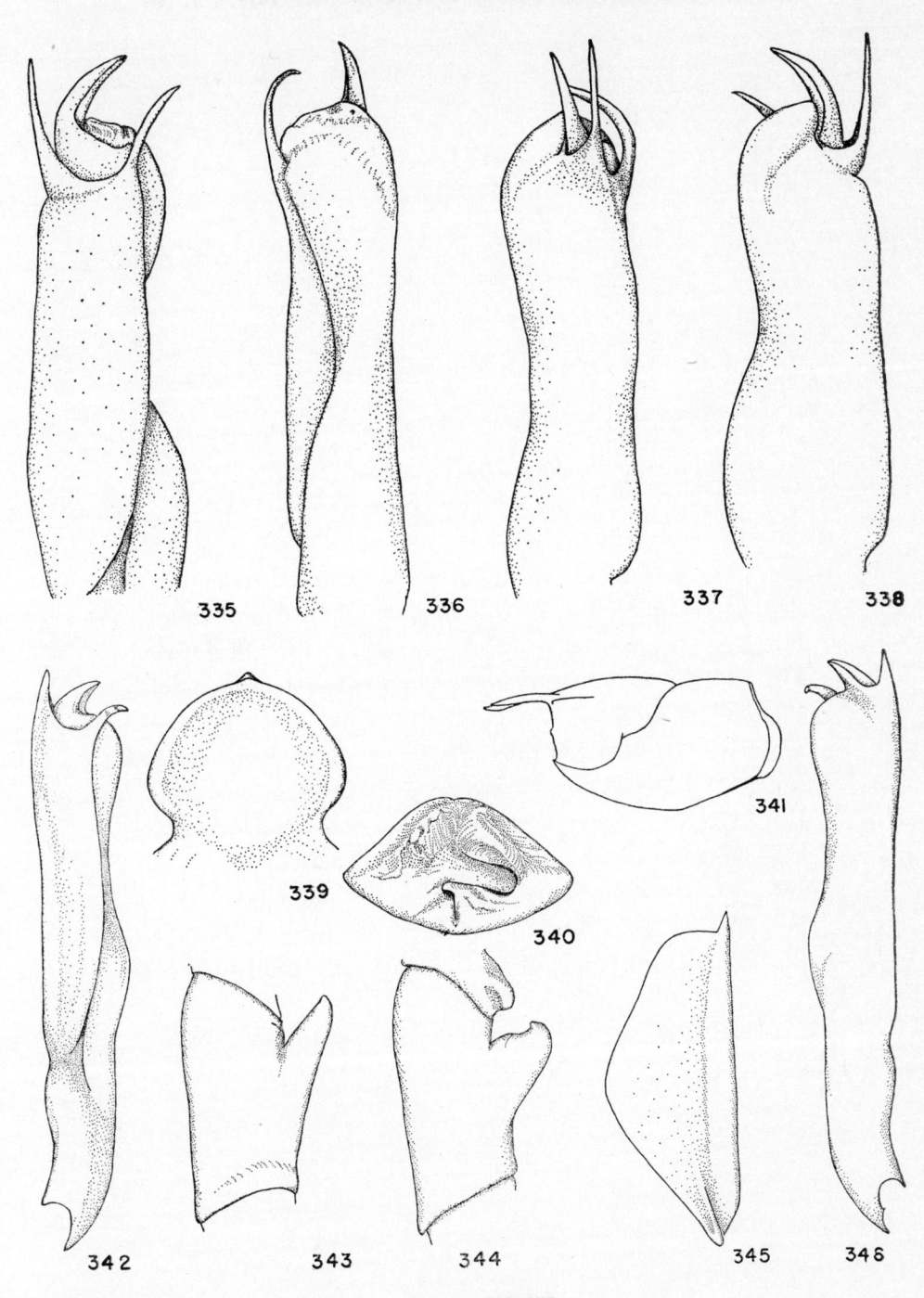

Plate XXIV